Approaching Machine Learning Problems in Computational Fluid Dynamics and Computer Aided Engineering Applications

A Monograph for Beginners

Justin Hodges, PhD

DEDICATION

To Clara, may my small achievements inspire your own wondrous journey. Dream boldly, for your potential outshines us all. I wrote the majority of this book while watching over your sleep with the baby monitor sitting turned on and next to my computer. What a special memory and blessing you are every single day. I love you endlessly.

ABOUT THE AUTHOR

My name is Justin Hodges and I am the senior AI/ML technical specialist at Siemens Digital Industries for our Simcenter portfolio. I sit in a product management role for our portfolio of simulation software. I earned my bachelors, masters, and PhD in mechanical engineering at the University of Central Florida in thermofluids. I grew up in a Siemens Energy turbomachinery lab characterizing heat transfer, fluid mechanics, and turbulence in gas turbine secondary flow systems. My masters and doctoral research was on film cooling flow-fields, predicting turbulence and thermal fields with advanced turbulence modeling and machine learning approaches.

During an internship in Princeton NJ in 2017, I fell in love with AI and have been building my technical proficiency with such since. In that time, we patented work which incorporated CFD flow-field results, based on CT scans, into machine learning models. Incorporating information from clinical surveys, (physical) breathing measurements, and CFD results into a machine learning framework provided improved diagnosis of COPD.

From then until now, I have made constant professional efforts to augment the capabilities of engineering simulation by virtue of machine learning. I help real customers solve relevant problems which add value to their business and delivered products. Whether it be more convenient or faster workflows, more accurate simulations, or better connectedness of multi-disciplinary teams and data, it is my passion to expand the usage of machine learning for the benefits of those in industry.

CONTENTS

1.   THE FOCUS AREA OF THIS BOOK

In this book we will strive to better equip the reader with tools and knowledge for when they are approaching machine learning (ML) projects – and not just any project, but specifically CFD/CAE based applications of machine learning. Generally speaking, this means there is prospective data which may already be generated (or planning to be generated) and the objective is to accurately fit a machine learning model to it (bonus if it's interpretable and/or computationally inexpensive). We're not really focusing on things herein like pioneering the development of a novel algorithm/architecture or an AI-based solver, or a large language model.

Without spoiling the subsequent sections, I will just state that we want to often do 'some specific steps' in our machine learning projects from as we complete them from start-to-finish, as well as additional steps to analyze and manipulate the data to maximally improve our resulting machine learning model and its usefulness. This book focuses on those steps specifically, which can be used as a template to complete future machine learning projects. The goal is to leave the reader with a sense of what the steps are in a typical machine learning project and how to execute them in your projects. I hope you feel empowered with this framework of thinking to approach future machine learning projects with confidence.

Here are a few example projects you could find yourself working on that would be relevant to the content in this book.

- Adjacent design characterization with ML-based reduced order models (ROMs) and historical datasets. In short, train a ML model to predict designs/scenarios you don't want to simulate or test.

- Robustness, Reliability, and Sensitivity analysis with ML models, perhaps following an optimization study.

- Quick and 'directional' results predictions to gain insight into a design space prior to running simulations.

- 'Big data' engineering projects (extremely large datasets that are too complex to be effectively processed, analyzed, or interpreted using traditional data processing methods).

- Integration with other technologies: ML can be integrated with other technologies like Internet of Things (IoT) and big data analytics, enhancing the overall capabilities of simulation.

- Developing new simulation methods which incorporate ML technologies.

- Incorporating data from different disciplines/teams/lifecycle stages to make better model predictions.

- More useful Digital Twins (DTs) and executable Digital Twins (xDT).
- Supporting 'the operator persona' with real-time insight made possible by ML models (e.g. dashboards to view higher fidelity results from field measurements).

- Replacing empirical models with more well-tailored ML models.

- Using ML models to reduce the computational cost of specific blocks in our numerical approaches.

- Inverse design/reverse engineering via ML-based models (feeding the generative AI theme).

- Real-time monitoring and control: In industries where fluid flow is critical, ML algorithms can offer real-time monitoring and control.

In case you want more specific examples, refer to the "DATASETS & PROJECTS" chapter for a more exhaustive list.

As you can tell from the above list, many of the use cases involve exploiting a machine learning model as a form of a ROM or surrogate to be used to achieve some speed-up or increase in fidelity. The idea of regression models, making and using correlations, and statistics is of course not new for mechanical and aerospace engineers, and may resemble similar tasks to a data scientist working in computer-aided engineering (CAE) applications (and many other industries). In fact, the 2021 Kaggle Machine Learning & Data Science Survey listed linear regression as the #1 most commonly used model in industry by machine learning and data science professionals. So, how do these things, some old and some new, relate?

- **Statistical models:** These models, like polynomial regression and linear regression, are typically driven/applied through domain knowledge. They often assume a specific form for the relationship between the independent and dependent variables (which could be from domain knowledge). When fitting a model to your dataset, your degrees of freedom are usually very limited to things like coefficients in a fixed mathematical form (e.g. linear). One example from boundary layer theory; linear behavior in the viscous sublayer and a log-law form at further wall distances. Another example, especially in non-Newtonian fluid mechanics, can be described using power-law relationships. So, if you opt to apply statistical models to your data, then you could often find yourself assuming the form of the data from your physical knowledge and then using the coefficients which improve the fit as much as possible.

- **Machine Learning models:** These models don't necessarily assume a specific form for the data. They're more data-driven, trying to learn the relationship directly from the data with more flexibility in the proposed form. Techniques like neural networks, decision trees, and support vector machines fall under this category. While similar, the form is significantly more open to change than in more straightforward and traditional regression models. Additionally, these models can be made significantly more complex, but usually as a trade-off to interpretability. Some specific avenues for customization would be changes to the ML model's architecture. For instance, a neural network can have multiple layers, each with varying numbers of nodes, and use different activation functions. Keep in mind that we could add other aspects like dropouts (for regularization), batch normalization layers, skip connections (like a residual neural network (ResNet) architecture) and more to the architecture, offering us significant options for modifying the model's form.

For fun, we can also extend the comparisons to the Buckingham PI method and feature engineering. In the Buckinham PI method, we essentially want to provide all the variables at hand in our project and use the method to realize certain advantageous outcomes: reducing the number of variables, highlighting important parameters and yielding a better understanding of their relationships, and sometimes benefits from introducing scaling/similarity-based transformations. If you know a bit about data science, this should sound familiar! These are all great benefits we could hope to achieve in our machine learning project via feature engineering, data pre-processing, and data transformations, among other things.

Again, recall this book aims more towards being a monograph for the simulation expert who wants to increase their machine learning literacy. Therefore, we will remain concentrated on this focus area.

This book might be relevant for you if you have similar questions to these in the focus area of CAE/CFD applications:

- What are some specific challenges and best practices in applying ML in CFD projects?
- How can model results be interpreted and used?
- How can I iteratively make changes to my ML models to improve it?
- How do I handle inconsistent, missing, or skewed data for my machine learning project?
- How do we judge model performance for a variety of datasets, or in a vast design space?

2. SETTING UP YOUR MACHINE LEARNING PROJECT

A machine learning pipeline refers to the sequence of steps or processes involved in developing and applying a machine learning model. It encompasses the entire workflow, from data collection and preprocessing to model training, evaluation, and usage. It's a systematic way of handling the entire workflow of a machine learning project. The pipeline ensures that the machine learning process is organized and reproducible, allowing for efficient development (and maintenance of models, but that's not covered so much in this text).

In my experience, knowledge on these topics is often sorely lacking in teams of simulation and testing experts, typically comprised of mechanical and aerospace engineers. Large enterprises may have dedicated data scientists that can work collaboratively with them, but small and medium sized organizations would not regularly have the same luxury. It is important that application experts possess knowledge of such machine learning steps to use in their projects because one of the most valuable things to integrate in your pipeline is domain specific expertise.

The pipeline and sequence of steps will not be the same in every project, but a general framework is proposed in Figure 1. Let's start with some clerical points on organizing a machine learning project, and then get into explanation of each of these blocks.

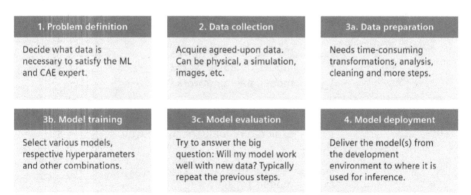

Figure 1: A general set of steps to define a machine learning pipeline. Letters are used for steps 3a-b-c to represent the iterative nature often needed to repeat these steps for different experiments.

Just like any project, organization can be very helpful and especially so when the project gets more complex, goes for longer timelines, and especially when the team involved grows larger in headcount. Machine learning projects are no different, and ideally, we set things up to be 'plug and play' so any changes, new models, or new experiments we embark on will be organized and require minimal changes. This is easy to say, but hard to master. As one example, as you refine and train more models to get better results you may consider tracking the progress for different models using git (a popular control system designed to track changes in source code during software development). Towards the end of your machine learning project, you may be running many different optimization/design exploration experiments in order to squeeze out a few last percentage points of accuracy in your 'final' model. If in a team of more than two people conducting these experiments, organization will be of paramount importance to ensure the time and compute is efficiently leveraged while fine-tuning the models.

There will be several files to possibly juggle in a project: the different versions of the data (raw, as well as variations from pre-processing), scripts for dividing up the dataset into sub-partitions, scripts for training models, scripts for making inference with different models (e.g. folds), different models you want to save after preparing, scripts for post-processing, among other things, to name a few.

Here is a brief visual aid I can provide to describe my baseline preference for a starting point on how to organize my project. Figure 2 is looking inside the contents of a sample project folder.

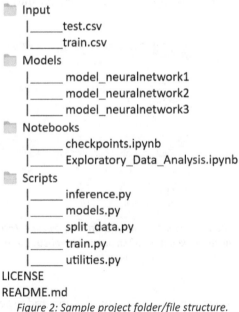

```
Input
    |_____test.csv
    |_____train.csv
Models
    |_____ model_neuralnetwork1
    |_____ model_neuralnetwork2
    |_____ model_neuralnetwork3
Notebooks
    |_____ checkpoints.ipynb
    |_____ Exploratory_Data_Analysis.ipynb
Scripts
    |_____ inference.py
    |_____ models.py
    |_____ split_data.py
    |_____ train.py
    |_____ utilities.py
LICENSE
README.md
```

Figure 2: Sample project folder/file structure.

Let's go one folder at a time and provide a brief description of each.

Input. This will contain the data used as input to the project, which will mostly be the data used for training and evaluating the ML model (but can include other things, like descriptions or supplemental information). If you have ran one hundred simulations for example, then you might have 100 subfolders with the respective data files for each. These files could be images, cgns files, csv files, or plenty of other options. At the first level under the Input folder, you could often have a single overview table file (.csv or .txt) which provides a summary of all the cases ran/generated to overview the dataset. Specifically, if your independent variables and outputs are 0D (single scalar values), then you can summarize the values for each variable (columns) for each of the samples (rows). If you have other results (e.g. 2D contours) that correspond with the 0D values for each sample, then you will need to include both into your machine learning models. The linkage between 'meta data' (tabular overview information) and the tensors/point clouds (e.g. 2D or 3D results) can be made with .json files, .geo and Ensight files, or hardcoded manually.

Models. Straightforwardly enough, this is where we store the machine learning models after training. Models can take a substantial amount of time to train, and money (e.g. paying for compute time on AWS), so it's advantageous to save them after training and import them in the future when doing experiments. You'll also probably have files associated to the training (e.g. events files from TensorBoard), so you may consider saving these files (for one each training event) in the same location so you can review the convergence behavior for each respective model. If you run a model search/optimization study, you could even store thousands of models throughout.

Notebooks & Scripts. These two folders may not be sensible for your project, depending on your setup, but is a structure often used. My preference is to run *.py files as scripts (without a GUI) so that they can run in the background (or from a Jupyter Notebook if I am more incrementally doing experiments in a very hands-on way). You may often find yourself executing these in a terminal or command prompt on high performance computing (HPCs) clusters where Python is installed, especially once you connect to a GPU node for more intensive calculations (e.g. 'ssh name@gpu.address.here'). Often, there will need to be utility functions/scripts ran upfront to prepare the dataset for training or other tasks during your machine learning project, and running these .py files in batch is my preferred style. Conversely, I like to use Jupyter Notebooks for conducting interactive and dynamics tasks (e.g. data/results analysis). One example is if I am trying to explore distributions for my dataset sampling, or if I want to try plotting my results in unique ways to better understand my model results, then I would execute this in the 'cells' of the notebook. Keep in mind you can also run some of the scripts from the notebook (for example with syntax '!this-is-a-file.py'). So in short, Jupyter Notebooks can be used for interactively working with your data (e.g. exploratory data analysis) and .py files can be used for automatic runs of certain steps in your pipeline without a GUI.

The last comment I would like to make is for managing projects as the number of experiments starts to become substantial. This can be difficult to keep up with as you run many (dozens) of experiments, as you'll be generating a lot of data for each single trial that needs to be kept organized: the hyperparameters used for that case, the produced results, the convergence behavior, the cross-validation used, and the manipulations/transformations applied to the data you used, to name a few. There is no one-size-fits all, but I quite like TensorBoard for keeping information on the models behavior (convergence, loss, accuracy, model graph, weights and biases, and more) which can be files saved locally ('events') and then opened later. To help us with this need to keeping track of many different experiments, I like the Weights & Biases MLOps platform. You can do a lot with Weights & Biases, like tracking experiments, versions, iterations of the dataset, reproduce models, and more.

There are many more considerations to be had as you imagine a production environment, further once your team grows to several people, but this outline is a reasonable start for running projects as an individual getting started.

When setting up your project, you may also benefit from a short technology stack review of commonly used tools in CFD/ML projects. Let's get into it below.

Linux operating systems
Like the CFD industries, a common high-performance compute (HPC) resource is a Linux-based operating system kernel that could be in the form of a desktop machine, a data center (e.g. cloud service), or server. Both ML and CFD libraries are commonly developed to run on Linux, so it's a good idea to brush up on command line (terminal) syntax and operations in order to manage your files, run your jobs, and call different software (e.g. python) to run.

Git, GitHub, GitLab
Git, in general, is a distributed version control system. It is used to track changes in source code during software development and enables multiple developers to work on the same project simultaneously without conflicting with each other's changes. In the case of a machine learning project for example, the project files would be organized in a repository (repo) and every time a file is changed from within such repo the changes are documented in a history ledger. This can allow users to jump back to older versions of files/code, which is nice because you can compare different states and do experiments on each. For GitHub and GitLab

specifically, they are similar in that they are both web-based platforms that offer hosting services for Git repositories (you can google GitHub right now and immediately start browsing through public code projects). They provide features to track issues, do code reviews, and generally facilitate collaboration for team members in effort to streamline development projects. While GitHub is mostly known for community focus and inclusivity, as most readers will probably have visited GitHub browsed around before, GitLab is slightly more focused on being useful for DevOps people. What does this mean? It means they focus on more of the union of software developers and IT professionals for collaborative efforts to deliver applications and services at high velocity for their corporation. This can include themes of development, deployment, automation of processes, monitoring and feedback, improving quality and customer satisfaction, to name a few.

Python (programming language)

It's hard to argue that Python is not the number one programming language for machine learning. Python code is operated on in a line-by-line manner by the interpreter and is one of the easier languages to learn for many, due to the simple, readable syntax that is helpful for writing and understanding code. The vast popularity of Python over the years has led to many machine learning libraries being built specifically for Python, making it a natural choice for your next machine learning project. Why? Well, libraries simplify and expedite the process of developing machine learning models, so you can use pre-built functions and conveniences to accomplish your machine learning project rather than spend a great deal of effort to code things from scratch. Some of the most well-known machine learning libraries you can use in Python would be TensorFlow, PyTorch, Scikit-learn, Pandas, NumPy (generic), and Matplotlib (generic), among many others. I consider myself purely mediocre in Python, so I will stop here after one more note to help it all make sense when you hear different phrases as a beginner: Python is a programming language, which you can code in with various choices of integrated development environments (Spyder, PyCharm, and Visual Studio Code are some of my favorites) to execute your machine learning project conveniently through the help of installing (use 'pip' command) different machine learning libraries (e.g. PyTorch). An integrated development environment (IDE) is like a tool, whereby Python is the material. An IDE is a software application providing comprehensive facilities to computer programmers for software development, while Python is a high-level, interpreted programming language. Lastly, while these IDEs are very popular, I would be amiss to not mention again that Jupyter Notebooks are also very common and purposeful in these projects, as mentioned above in a few case scenarios.

ParaView

While some commercial products have really nice visualization capabilities, open-source tools are awesome too. ParaView is well renown for a go-to visualization tool with scientific data. It is scalable in processing big data packages, whether be from single-processor machines to distributed parallel computing environments. It is well known in handling mesh-based data with its visualization toolkit (VTK). One particularly attractive characteristic is that user's can use the graphical user interface (GUI) or script their visualization and analysis tasks using Python (batch operation). This flexibility allows for both exploratory analysis and automated workflows in a really repeatable and rapid manner, which can be key for industry timelines.

Honorable Mentions

Docker. Making containers is a great way to keep your projects reproducible while running on different machines. For example, if you want to share your AI framework code on the internet and have anybody download it and run without (many) issues. It packages ML models and their dependencies into containers, ensuring consistency and making deployment across different environments easier.

FastAPI. If you want to avoid running code from command line/terminal, and you want others to be able to connect and use it, it's a great option for you to package it inside an API. FastAPI is a web framework to quickly build Python APIs. It's known for creating web services that can serve your machine learning models.

Python Libraries

NumPy. Provides support for large, multi-dimensional arrays and matrices, along with a vast collection of high-level mathematical functions.

Pandas. Offers data structures and tools for effective data manipulation and analysis, particularly for tabular data. It is similar to NumPy, since they are both for data analysis, differences being their intended use cases and the types of data they are designed to handle. Pandas provides specialized data structures and functions for more complex data manipulation and analysis, especially with tabular data, which is not as directly supported by NumPy's array-focused approach.

Matplotlib. A plotting library for creating static, animated, and interactive visualizations in Python. It's not machine learning specific, but you will use it often in your projects.

PyVista. A Python library for 3D visualization, which is especially useful for CAE data (and more generally scientific data). Whenever your visualizations are complex and you want more interactive plots, especially since we find ourselves working with spatial data quite often, it is a good option to reach for.

Scikit-learn. Focuses on machine learning, offering tools for data mining and analysis, including various algorithms.

TensorFlow. A library for machine learning and artificial intelligence, particularly known for its capabilities in deep learning.

PyTorch. PyTorch offers dynamic computation graphs that facilitate deep learning models' development and training, making it particularly suited for research and flexible model design.

Keras. Provides a high-level, user-friendly API for building and training deep learning models, making it accessible for beginners and efficient for rapid prototyping, distinct from TensorFlow's comprehensive platform and PyTorch's research-oriented flexibility.

3. PROBLEM DEFINITION AND CAE THINKING

The first step is the definition of the problem. This is almost a definition of two problems: what are the actual physics and domain knowledge to focus on when aspiring to make a machine learning surrogate for a simulation project, and secondly the machine learning approach/data that is needed to satisfy such project goal. This would require collaborative discussions and planning between those with machine learning knowledge and those with simulation/domain expertise (unless any individuals possess both). Generally, since this process can vary from project-to-project and with different industries, a services-based approach is quite common among the broader tech industry for AI companies to scale their business and manage adoption of AI at new companies. The connection of the business problem, which in this case of a CAE setting also requires technical expertise in the principal domain (e.g. aerodynamics), to the data science side is quite time consuming as both parties need to exchange expertise with one another until a mutual understanding of each side is reached. This is one of the most critical stages in the project for several reasons.

Garbage in means you should expect garbage out. The problem needs to be posed in a meaningful way or the outcome (ML model, predictions, etc.) are useless. For example, if the goal of a project is to create a model that can learn the aerodynamic behavior inside a turbomachine, the appropriate independent variables must be selected which are relevant and influential on the resulting flow-field. Sensible choices could be the devices rotational speed, the flow angle of incoming flow, the turbulent quantities which describe the flow-field, the roughness of the surface finishes, etc. Further, an important exercise is to consider the range of values for each parameter and how that will affect the resulting aerodynamic performance. If certain aerodynamic phenomenon cannot be observed within the parameter space that was sampled when creating the dataset for the intended purpose of making a ML model, then it should be very explicitly highlighted that such a machine learning model should not be used for inference in such a case outside the bounds of the data whereby new phenomenon may occur. In fewer words, this reflects one of the most common questions I receive in the field: can my model be used for more than just interpolation in a design space after it's trained. In the turbomachinery aerodynamics example, a concrete scenario could be using this ML model to predict the stall cells (separation and recirculation zones) in a compressor flow-field, despite that the majority of the cases used for training did not possess such stall behavior. Rather, if stall prediction was the main intent for the machine learning model, then the sampling of the input and output variables should cover various types of behavior (e.g. small cells, large cells, cells in the different locations, etc.) so that similar phenomenon are present in both training and inference. By the way, the term 'inference' is often used to describe generating a ML model prediction after it's trained (at least that's how I will use it in this text). So you can think about training as one phase, and then using the model afterwards for inference to make predictions in new unseen cases.

These discussion points in the problem definition stage create more questions to resolve. For example, using a ML model for interpolation versus extrapolation, how to make use of historical data to broaden the predictive capabilities of the ML model, and as such what ML methods are most appropriate for that type/size of data. Further, it's worth noting that while you have probably heard of elegant physics-based ML algorithms that naturally have (somewhat) interpretable properties, this type of simple and conceptual thinking in the problem definition phase is one effective way to add more interpretation qualities to many ML algorithms. This is very helpful during validation and evaluation stages after your model is trained and you are trying to assess how you can safely apply it. This is a great benefit to the simulation engineers most times in adopting such ML methods.

Hodges et al. [1], among countless others, demonstrated some of these concepts in a CFD study for an internal combustion engine that focused on the influence of the geometry on the resulting mixing/combustion. Both the geometric variables that were inputs to the study and a sample contour plot through the domain of turbulent kinetic energy can be seen in Figure 3. From the inlet flow angle, labeled as "1", the flow separation will be strongly affected by choice of a shallow or steep angle. Considering the objective of having a machine learning model that can predict the mixing characteristics within the chamber, the choice of which angle values to train/test/validate on are critical, as the flow-field will be foundationally different for different ranges of angles. By understanding the taxonomy of what the secondary flow structures look like in the training samples one can have more physical understanding of the limitations of their machine learning models. This approach for digesting and formulating the problem statement is helpful for the simulation engineer to retain confidence on when and how to use the model in their design work. As you can tell, it requires domain knowledge and cannot be realized alone by data scientists.

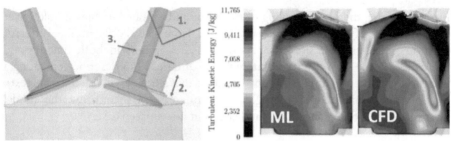

Figure 3: Geometry and sample results for an internal combustion engine study with CFD and machine learning [1].

As alluded to previously, simulation practitioners in CAE have an added layer of challenge in their problem definition process, given the challenge in understanding the physical behavior present in their simulation results which originate from non-linear partial differential governing equations. This understanding is ideally acquired before sampling is done for preparing the dataset. Next, let's spend some time to try to bridge the gap between typical simulation data (which the reader is probably quite comfortable with) and the transformation of such into our machine learning dataset. To draw on familiar concepts, we will draw metaphors on concepts that are common to mechanical and aerospace engineering curriculums in order to introduce the reader to aspects of machine learning. As a fun irony, I will point out that this concept is used to enhance the capabilities of machine learning models and is dubbed 'transfer learning'. So the irony is in full effect; teaching humans ML by the same methods ML models learn from human data!

As we begin with the metaphors, I will just point out that none of these are factually or literally correct when articulating what machine learning is like. We are not literally defining machine learning as the following subsequent items. Rather, these are (hopefully) just great concepts to open with to get the reader familiar with a framework of thinking which machine learning resembles when framing a problem and matching with subsequent data. Whether it be a university lecture, a training given to colleagues, or engagements with a customer, I have had very positive feedback from beginners on the usefulness in these illustrations to help get them started.

To begin, we can oversimplify and compare a machine learning model as analogous to that of a closed-loop transfer function. From personal experience, I took an undergrade coursework in feedback controls course for dynamic systems and learned about control theory and transfer functions at a surface level. In hindsight, this depiction of a response (an input signal, an output signal, and 'everything in between') at least conceptually resembles the way we use machine learning models to provide a linkage between our outputs (responses) and our input variables (input signal). Please note that I am not equating the mathematics whatsoever between transfer functions and machine learning models, but rather only showing similarities in a 'flow chart' type way for how the data is created and then applied when using the ML model as means for creating a correlation. Rather than the term on the right-hand side of the equation in Figure 4, a neural network would have a different formulation that is based on everything between the input layer and the output layer in the model architecture (how we refer to the layout of the network).

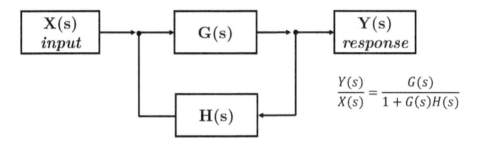

Figure 4: An example of a closed-loop transfer function (equation included) [2].

Another familiar concept to draw on and compare to is that of linear regression. Which in itself is debatably a machine learning technique; however, linear and logistic regression were found to be the single most popular method & algorithm used from the Kaggle State of Data Science and Machine Learning 2021 survey [3]. My personal opinion is that it ranked so highly in popularity because it is a common technique to use early on in machine learning studies, especially for rapid analysis with interpretability which is especially helpful in the beginning phases of a project, not because it is a class of machine learning techniques in pure form. Even being linear, it is still useful to fit models with a high number of dimensions (for example a 15-parameter space), since humans really struggle visualizing anything beyond three. But for now, we can ignore the debate on if it is included in the machine learning family and just build from it as a familiar concept we have probably learned already in our training.

In linear regression, a finite set of data samples form a trend which a linear function tries to accurately model. The data, in this scenario as an example, is described as simple linear regression where one variable y is dependent on one variable x, rather than y depending on multiple variables and thus having multiple dimensions. The linear model has parameters which are initially unknown, but the provided data is used to guide and set the parameter values. The task of linear regression is to use the provided data to come up with parameter values that minimize the error (distance) between each point and the proposed linear trendline. An example of simple linear regression can be seen in Figure 5 [4].

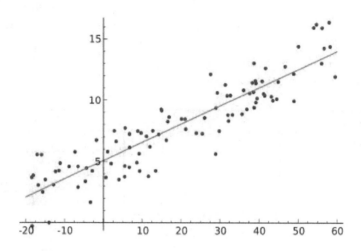

Figure 5: Example of simple linear regression with data points and trendline shown [4].

This is very similar to our task at hand when fitting machine learning models; only the complication in both our challenge and the methods needed to solve them can go up significantly! Even imagining linear relationships in more dimensions quickly becomes impossible (2D, 3D, 4D, …). Let's consider stepping up the problem complexity and a slight pivot from a regression to a classification problem. In this case, let's pretend we are trying to discover a curve that accurately splits data between several classes (like a label). In the linear regression figure, that would imply all samples above the line receive one classification and all below the line receive another. We can use the famous Modified National Institute of Standards and Technology (MNIST) dataset [5], which is comprised of many images of handwritten number digits and commonly used for training and testing image classification algorithms. From a recent work by Wach et al., one of their figures provides a nice visualization to increasing complex and non-linear decision boundaries between multiple classes. The Figure 6 shows how different classes (digits) can be segregated from one another in a dataset, from left to right, as more classes are subsequently included for the model to learn. These data samples look like they form well segregated clusters (black markers) for each class at first, but when just five different digits are considered, you can observe that the decision boundaries are challenged to precisely subdivide and capture all samples for each respective class (example: red region for digit '3'). This is just one conceptual example to build from the familiar concept of linear regression. You may also be realizing that the ability for machine learning models to manage highly non-linear and high dimensional data is at the core value-add to society from previous methods, and in part cause for its immense success.

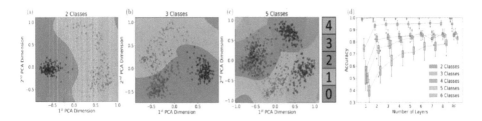

Figure 6: Depiction of different decision boundaries for the MNIST dataset [6].

Next, we can conclude the series of simple and familiar examples with a final exercise of comparing machine learning to a survey questionnaire. While machine learning models and questionnaire surveys serve different purposes and operate differently, there are some similarities between them which can be educational for someone starting to learn ML. One similarity being there exists a choice for the survey architect on which questions to ask in a survey that are most relevant, or equivalently in the context of a dataset what data is most relevant to include when building a machine learning model. Often there are input variables that we know have a first order contribution to the result (output), but unfortunately there are also input variables that do not have a clear effect on the output across a wide breadth of circumstances (parameter ranges). For example, Figure 7 considers a machine learning model that could be made to decide if a given set of design parameters and/or operating conditions would produce a 'good' aerodynamic performance. Some variables will have a high impact, like Reynolds number, but others may or may not have a significant influence on the aerodynamic performance, like the alloy type. Of course, this depends on the specific flow and thermal configuration, like a flow motivated by buoyancy effects, and could have a large impact on the aerodynamics. As is, this is a classification problem, but could easily be posed in the form of regression by being more specific on what constitutes a good versus a bad design (e.g. a classification problem could predict if efficiency is above a certain value or not, while a regression problem could pose the model to predict the efficiency value).

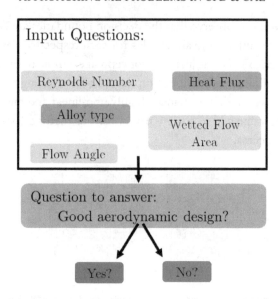

Figure 7: A "survey questionnaire" analogy for formulating a machine learning dataset. In this case, for variables relating to aerodynamics performance.

The survey questionnaire template way of thinking also matches well with the structure of a tabular dataset, which in this case could be a table compiling information from a batch of individual CFD simulations summarized in a single table. For this case, our inputs describe the respective simulations and would be entered into the table column-wise. Whereby the simulation results, for example the aerodynamic efficiency, would also be entered into the table column-wise. Lastly, each individual simulation would be a separate row (we could also refer to it as a sample). This is certainly not the 'only way' a CFD dataset can be formulated for a machine learning study, but it is a common one used in an industrial setting and a very good basic one to review for educational purposes. Let's visualize this structure, starting with Figure 8. I have found this discussion really helps break down people's apprehension in going from "I have my simulation study results done" to "I have my machine learning dataset formatted properly".

	Re	AoA	RPM	...	η	ΔP	...
1							
2							
3							
.							
.							
.							
N							

Figure 8: Sample table from a CFD study comprised of N individual simulations and $F_1 + F_2$ columns (inputs + outputs).

This table of data summarizing a set of simulations can often be stored into an excel file, a Comma-Separated Values (.csv) file, or other similar formats. An important consideration is the shape (like an aspect ratio) of your table. As more columns (inputs) are added, the capability to fit a correlation with the same number of samples (rows) tends to be more difficult and thus require more data. There is no absolute rule of thumb for all scenarios on how much data is needed as a function of how many variables are considered, because the relationship of the inputs and outputs can be simple (linear) or by contrast highly complex (non-linear) and thus have widely different data requirements. However, it is important to be mindful of the shape of your table in general while experimenting with your model trainings. This will allow you to be practical and confident in deciding how limited you are without adding more data. Generally, we want many rows with fewer columns (tall) instead of the opposite; many columns with few rows. There is no absolute rule, but many times in my experience there is a strong incentive (compute time and cost) to reduce the number of simulations used in the dataset, which will encourage your datasets to be short but wide in nature. There are lots of measures in AI/ML literature to deal with small datasets and they won't be covered herein extensively. When forced to quantify the number of samples required, I usually offer two pieces of advice:

1. Use an adaptive sampling approach while generating your data. This ensures you sparingly and efficiently use your samples for your specific project goal, usually with an effort to spatially fill your parameter space and/or add more samples to areas in the parameter space with steeper gradients. There are algorithms you can use to automate this process, such as active learning.
2. For simulation studies that end up in a summary table, meaning you want to fit a model to tabular data, I recommend at least as many samples as twice the number of columns plus fifty.

To finish these analogies before moving on, we will tie the resemblance of the survey questionnaires to a simple neural network via two illustrations. 'Simple' meaning a feed-forward neural network, which signifies that information moves in one direction, from input to output, without any loops or feedback connections. Keep in mind that the goal of this section is to write in such a way that a beginner feels clearer on how familiar types of data and datasets relate to machine learning models and pipelines. Comparisons to things like transfer functions and questionnaires are only meant to be conceptual, not to imply that the mathematical approaches or formulations are equivalent.

From the table in Figure 8 , we can map the parameters to the a simple neural network, as in Figure 9. 'Tunable parameters' here has a dual meaning; the permutations possible for different combinations of hyperparameters (e.g. number of layers) iteratively tuned to increase the model accuracy, as well as the learnable components of a neural network that are adjusted during training to minimize the error or loss function (e.g. weights, biases). The former is user specified in the setup, while the latter is performed automatically during the machine learning model training by the library you use (e.g. PyTorch). While the columns have a clear translation from Figure 8 to Figure 9, the rows may not be as obvious. Since the rows represent the samples, or in this demonstration example the number of simulations, they can be split into at least two ways. First, they can be divided into different buckets for training, testing, and validation. These will be covered in detail later in the text. Second, groups of rows can be aggregated and used as distinct 'batches' when conducting the model training, since the learnable components of the network (e.g. weights and biases) are calculated iteratively many times throughout the training processes on different batches of the data. The selection of how many rows included in each batch is also a hyperparameter called 'batch size', which is defined more generally as the number of training examples processed at once in one batch in one iteration of the training algorithm.

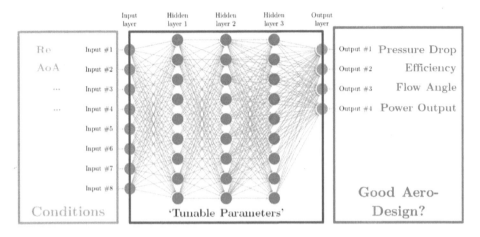

Figure 9: Posing the sample variables in Figure 7 and Figure 8 onto a simple neural network.

For the sake of completeness to connect this back to Figure 4, the green box in Figure 9 can be associated to 'X', the red to 'Y', and the blue to everywhere in-between. X and Y. While the mathematical processes between a neural network and the transfer function equation are completely dissimilar, the core of the analogy is that the whole purpose is to establish an accurate mapping of the inputs and outputs.

My hope is that this section has helped the reader form a better understanding of a few basic concepts:

- The challenge in knowing which variables are most relevant to the output result, as it relates to both the domain knowledge and sensitivities in creating a machine learning model

- The similarity between large survey questionnaires and machine learning. A large survey asks many peopley (rows) a series of questions that are descriptive and hopefully correlating to a desired outcome (columns). The parallel for simulation data being that many simulations can be run (rows) with different values for the geometric design and/or operating conditions (columns) which hopefully correlate to the result the model is trying to predict (columns).

- The concept of 'the opinions of many average experienced people' tend to outweigh the 'opinions of a few experts', which is a core idea in tree-based algorithms (random forests) as well as large survey exercises.

- The 'flow' of data (conceptually) from the output of a CFD study to the machine learning model. This includes how data can be compiled into a table, where these rows/columns originate from in the design space, and then how each variable in the dataset corresponds to different places in a neural network 'setup'.

4. DATA COLLECTION

As mechanical and aerospace engineers, we often work with a rich diversity of data during our design processes, which only becomes more lush after manufacturing when the system or component is in operation and analytics data can be gathered on how the hardware has performed over its lifestyle. Concepts like the Digital Twin place a high value on integrating both high quantities of data as well as different types of data into models. Integrating data of different types (for example CFD simulation data as well as experimental measurement data) can help to provide a more holistically accurate representation of the physical behavior, since the error from each of the individual sources (e.g. simulation error) is offset some by taking multiple sources of data. These models can go by several names, such as surrogates, reduced order models (ROMs), or for example executable digital twins, depending on the context. These terms are quite popular in the CAE community, but what may be less obvious is that a machine learning model could be the actual model under the hood taking in the consolidated data and producing accurate predictions (e.g. a neural network based ROM). I personally expect that this will continue to be more popular with the continued development and increased popularity of advanced machine learning models.

We can provide a short survey to explain different data types commonly used in the CAE domain which may find their into a machine learning pipeline. Firstly, let's begin with tabular data, which admittedly would be a short section since that was overviewed above already for a primary use case of summarizing batches of CFD simulations. Generally, and not specific to CAE, tabular data is one of the most common and popular types of data used in machine learning studies. Tabular data is a structured data that's organized into rows and columns, where each column represents a specific feature or variable, and each row corresponds to an individual data point or sample. Machine learning algorithms and techniques designed for tabular data, such as decision trees, random forests, support vector machines (SVM), and deep learning models like fully connected neural networks, are well-established and widely used.

In my experience, one of the most common formats of data that we work with in CAE for machine learning studies are .csv, .txt, excel, .dat, or similar. These are simple text-based formats used to store data and work really well for the aforementioned example of summarizing a bunch of independent simulations.

However, another popular use case for machine learning on CAE (i.e. simulation) data is to extract local scalar values on nodes/cells within your simulation (now our dataset has spatial characteristics). There are a variety of circumstances for which scalars in a solved simulation could be extracted and used for training a machine learning model; cell values on a given two-dimensional plane section, cell values over a boundary or surface, cell values in a three-dimensional point cloud, to name a few. By the way, 'cell values' refer to the results from the simulation, such for example temperature values at each cell in the simulation.

The nature of this data begins to make using simple .csv file formats difficult, especially as the size of the data grows (three-dimensional planes can have 10s of thousands of cells per sample, and three-dimensional point clouds could have 100s of thousands of data points per sample). Preparing these datasets after the simulations are ran requires pre-processing steps to (at minimum) organize the data prior to any feature engineering or ML-based transformations. You may find yourself spending a non-trivial amount of time to do these preparations, such for example, appending or joining tables, concatenating cases, associating meta data for each case to field values, etc. If you have a basic level of coding experience then these operations would be straightforward for you, but they still might take more time than you'd like. These are some of the operations that make a commercial no-code ML pipeline software suitable for applications like CAE; low barrier for entry and therefore very inclusive for people to use, as well as delivering a time savings with useful pre-canned functionality you would use often. With such limitations in text-based file formats as the datasets get larger and larger, one notable alternative is the use of CFD General Notation System (cgns) file formats, which are built to compile grid information, solution data, and meta data (flow variables, boundary conditions, etc.) for simulation runs. For this purpose, two notable formats which are similar are Hierarchical Data Format version 5 (hdf5) and Visualization Toolkit (vtk) formats, which are both commonly used for scientific visualization applications. As a heads-up, if you are extracting three-dimensional blocks of data from each simulation, for example velocity magnitude extracted over a range for x-y-z coordinates, it is very likely text-based formats like .csv would be insufficient.

Another popular format to represent spatial data is that of an image file (png, jpg). As in the case above for consolidating information from local scalar field values into machine learning models via tabular and/or cgns formats, an alternative is to take image snapshots of the displayed results and use those as the actual input data. A wide variety of examples for how these images can be created and used are evident in literature, but one example to use herein we simply consider taking an isometric view of a scene where the results are being plotted on the local cells over a no-slip wall [7]. In this case, we can see the piezometric pressure distribution on the cells over the marine hull body. These snapshots could be taken for each data sample, and associated to some 0D features (resulting drag, metrics describing the hull geometry, etc.). Some basic measures need to be taken when working with images like this, such as to ensure the same vantage point is used for capturing all images, remove annotations/text labels/colorbars/etc., use a transparent or white background, turn off ray tracing to remove unrealistic lighting, reflections, and shadows, and finally pay close attention to your resolution and window size (computer vision methods will have scaling limitations based on how many pixels each image has). While simple and intuitive, don't forget to do it upfront before preparing your models.

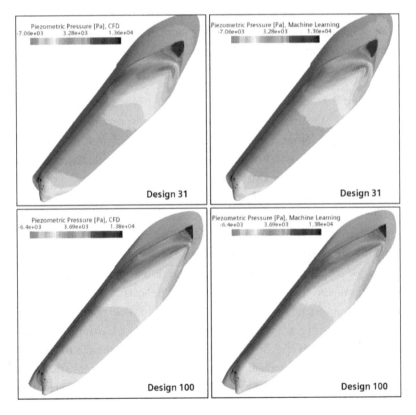

Figure 10: Sample image snapshot of a scalar distribution (pressure) over a no-slip wall (a

marine ship hull). There are four individual snapshots here collected into a grid to also show a comparison of the CFD and the machine learning results for two sample designs.

It should also be noted that there is a flexibility when working with images to convert into numerical data, in case that's convenient for you. First, you need to load the images (some popular Python libraries for this are OpenCV or PIL). Next you need to flatten the data into a 1D representation (array), which is usually done by extracting RGB color channel values for each pixel. Last, you may consider doing some feature scaling or normalization of the RGB values (some models benefit when the data is in a range of 0 to 1 or standardized with the mean and standard deviation). The choice of numerical or image-based data comes down to preference for different use cases in industry, as I have observed different teams that favor specific ones as a matter of preference for the way their design analysis pipelines are done.

Excitingly, when working with images one can leverage the massive amounts of research and innovations in computer vision from fields like autonomous driving and medical imaging. Just a few examples of powerful model breakthroughs in such fields, which I think have meaningful places in the CAE domain, are the U-Net models, ResNet models, VGGNet, InceptionNet, and EfficientNet for scaling up CNNs efficiently.

Another type of data that could be collected in a CAE application of machine learning is that of 3D data which describes the geometries/domain, such as CAD, STL, or other. When a machine learning model can import both the results and geometry data, it provides the benefit of rapid inference for different geometrical design configurations that may not be parameterized. For a parameterized design space, both the geometry and results information can be summarized in a table. Without a parameterized design space, the model must learn directly from the surface geometry/mesh, which is more complicated to do than providing geometric parameters. Variational Autoencoders (VAE) are one good technique which can learn geometries that don't possess any parameterization, and then even provide a machine learning based parameterization (a latent space). This is very useful for design engineers; more flexibility to work with raw geometries, the ability to provide dimensionality reduction (since you can specify the number of latent parameters to generate in your VAE study), ability to create new geometries from the latent space, and facilitation of optimization tasks by working in a lower-dimensional latent space.

An example below is shown from the software Monolith AI from the same work

32

as referenced above in Figure 10. The red and blue representations are overlays of both the original geometry (validation set) and the VAE model inference. There are nine pictured validation cases randomly selected, which are all unique hull geometries. In this case, the hull geometries were initially parameterized (which was later discarded), and such was used to sample 2,000 geometries. Note that this sampling entailed generating 2,000 geometry samples of the geometry, but not running any simulations of them. These were used to train the VAE (encoder and decoder) and re-represent the design space with the new latent parameterization.

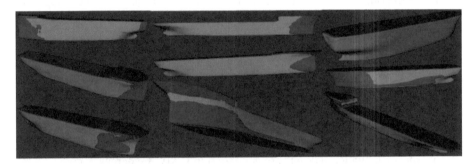

Figure 11: Using Monolith AI to learn the geometric representation of a marine hull with a variational autoencoder in a latent space.

Graph neural networks (GNNs) are also a fantastic model choice which excel at this task of learning geometries/spatial data, which have gained a lot of popularity and success in recent years. They are highly suitable for mesh-based data from simulations, since each node in the mesh can be considered as a vertex in a graph, and the edges (which are connections) between these nodes represent the edges in the graph. GNNs are also very favorable for industry simulations because they can handle unstructured meshes without issue, which is extremely common. More on this later in the section which covers various models.

After a brief mention earlier, let's provide more information on the topic of sampling. Sampling is critical to the resulting accuracy and usefulness of your machine learning model, and usually is an integral part of answers to difficult questions like 'how many cases do I need to run to train my model'. "Sampling" refers to the process of selecting a subset of data/cases from a larger dataset or design space, just the same as when doing design exploration studies with simulations. Let's scratch the surface with a small discussion on some different sampling techniques. Note that most of the time in this text we are referring to sampling as a means for picking which simulation/training data to generate and form the original dataset, not which data to segregate and use for training your machine learning model. The latter is possible, but most of the time the subdivision of the dataset for training/test/validation is randomly picked or by using domain knowledge to intentionally pick certain samples for experimentation (e.g. can my model train on high drag cases but yet generalize well enough to accurate predict cases with lower drag).

Random Sampling. Every individual sample or item in the population has an equal chance of being selected. It's akin to a lottery system. It's important to scrutinize the outcome of random sampling whenever dealing with very large or very small populations, as it might not be effective.

Latin Hypercube Sampling. The idea behind Latin Hypercube Sampling is to divide the range of each input parameter into non-overlapping intervals based on the desired sample size. It's a great approach when it's essential to efficiently explore (fill) the entire parameter space while generating minimal samples.

Optimization. Not ordinarily in a conversation about sampling approaches, but often a means by which datasets are generated in CAE. Generally, a cost function is formed (e.g. minimize pressure drop) and an algorithm seeks to iteratively select sample points in the space to generate (simulation) results that best satisfy the cost function. What's important to note for the context of this book is that while optimization datasets might be available for machine learning studies, they are often challenging to use and may need supplementation/pre-processing. Because the cost function is usually more sensitive to some variables than others, the dataset usually has some variables which are highly under sampled and thus problematic for a machine learning model to fit over wide ranges for such.

Adaptive Sampling. Initial samples are first generated to build a surrogate (e.g. machine learning model). Areas of interest (like regions with high sensitivity or gradients) are then identified, and more samples are added adaptively in these regions to improve the model accuracy. This is a very attractive option when you want to spend minimal samples (in case simulations are expensive to execute).

Now that a brief discourse has been done on different types of popular techniques to sample a parameter space, the discussion should turn to biases that can be present in your dataset. Bias is an important consideration, as it can affect the accuracy, fairness, and generalizability of the resulting models. I will list a few types of bias, but this shouldn't be taken as a complete list:

- *Measurement bias.* Occurs when the tools or methods used to collect data introduce error, leading to models that are based on faulty data. Many machine learning model projects with CAE data will often inherently use the simulation data as the ground truth, meaning that if the machine learning model can make predictions that fully match the simulation data it would be deemed 100% accurate. However, this is of course not the case and a misnomer, as simulations are also error prone and the physical phenomena being simulated are not always adequately captured. Incorporating other forms of data can help reduce this bias. Let's say we are doing a CFD simulation study for a flow over a cylinder validation case in the oil and gas industry; we could run several CFD codes from different vendors, we could incorporate physical measurements from pressure tapes or anemometry-based measurements, or additionally we could incorporate historical knowledge from things historical data from within the company or textbook correlations.

- *Selection bias.* There are several related terms for this one, such as sampling bias, representation bias, and probably others. Sampling bias can occur when inadequate measures are taken to create the dataset by sampling, which leads to a lack of diversity of the data points, with some trends and anomalies not captured. These could be trends in small pockets of the design space, behavior of edge cases, and/or complicated non-linear trends missed among some variables, among other things. To illustrate this, consider you are running a DoE study with mechanical FEA simulations based on a set of ten geometric parameters describing the solid body. Selection bias can become an issue if the samples of simulations selected are for tight ranges of the geometric parameters, relative to the full feasible range for each geometry parameter. In this case, there could be unique behavior for the mechanical results in said range of parameters that were not used in the DoE, and thus the behavior is not well captured by the model, since it was not exposed during training. To avoid selection bias, it's important to consider full possible ranges for the different configurations; both in terms of the training and the application of the model after created. This is another reason a structured ML pipeline is necessary- you can identify traits like this in your dataset and take steps (pre-processing) to fix them.

- *Experimenter Bias (Observer).* Simply put, different people may run simulations differently and those differences can cause an additional bias to the data when aggregating the simulation results from different authors into a broader dataset. Some practices regarding simulation setup are more of an art than science, like meshing or best practices when setting up solvers and physics models. While it is probably alarming to read that a major part of a scientific simulation can be labeled 'as an art', I think it can be acknowledged as a reality when you consider the substantial differences of opinion and practice in organizations that have large simulation departments. There are many relatable examples in the CFD industry, such as different preferences for turbulence models, particular versions of the same CFD code, different convergence criteria/tolerance a user would prefer to use, or different model settings in core solvers to balance between solver speed and stability, to name a few. This parallels precision bias in experimental measurement; how different people can use the same instrumentation and hardware but yet conduct their experiments differently, leading to different results.

5.   DATA PRERARATION AND ANALYSIS

After data collection, the next phase in a data science or machine learning pipeline may go by many names, but essentially represents an iterative process to analyze and manipulate the data prior to conducting model building and training. I am choosing to call it data preparation and analysis, but I struggled to settle on a name for this section because there are so many alternative names/steps. For example, data cleansing, data pre-processing, data extraction and analysis, among other variations. My naming convention was an attempt to encompass all of these into one category.

Exploration Data Analysis (EDA) is an important first step to identify general patterns in your data, which could be helpful or threatening to the ambition of successfully building and applying your machine learning model. It includes analysis, visualization, and exploration of your data in its most raw form after collection. Ultimately, EDA is about open-ended investigation into the data. It does not start with a predefined hypothesis but instead allows the investigator to ask questions, probe into the dataset, and uncover insights that could lead to a better resulting ML model.

Up to this point in the process the problem definition has been formed, which probably more heavily focused on the business logic and domain (e.g. physics) expertise than the machine learning considerations (e.g. is my data going to yield a good model). I view EDA as one of the early key steps to ensure that the machine learning problem is formulated with continuity to the problem domain. A different way to state this is that we want to maintain a physical understanding of the problem that we are applying machine learning to, and EDA allows us to maintain that physical understanding during the entire process of manipulating and preparing the data for model training.

For example, let's consider a machine learning problem focused on flow over a flat plate, where we use variables that we know things about from fluid mechanics (Reynolds number, heat transfer coefficient, and pressure). In our EDA, among many other steps, we would plot the relationships of different variables over their ranges from data collection, such as heat transfer coefficient over a wide range of Reynolds numbers. When we notice a rise in heat transfer coefficient when the Reynolds number transitions from laminar to turbulent, from a machine learning point of view we may consider segmenting this data into separate bins for all the different flow regimes, as interpolation of both laminar and turbulent data may cause poor predictions later by our model. Further, we could note the proportion of data we have collected from each regime, and possibly make transformations to our data if we were worried about an imbalance between each. These are all important considerations which can improve the outcome in the ML model. Because we know the basics of fluid mechanics and heat transfer, we can identify these necessary actions to take to manipulate the data. Imagine we have no sense of understanding in our dataset that some of these patterns, and problems, exist. Our hope is EDA could bring them to light so they could be remedied.

Let's use a case study on a specific dataset to illustrate some common steps in an EDA with code. The dataset is for a simple CFD study on a converging-diverging nozzle. The inputs are geometric parameters which describe the converging and diverging sections, while the outputs are the resulting pressure and velocity averages at the discharge (exit) of the diverging section on the right side of Figure 12.

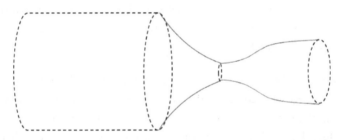

Figure 12: A sample diagram for a converging-diverging nozzle. The left hand side is a chamber attached to the converging section, and the right side shows the diverging section discharging into atmosphere.

We begin by importing the necessary libraries to perform our EDA as well as the dataset, which should be stored in the same location as the notebook/Python environment. Since this is the first code appearing in the book so far, I will reiterate that this book is meant to be a 'code along' book, meaning that the reader independently with the support of key sections of code from the book. All of the code won't be provided in this book, and you will need to fill in missing pieces.

```
#import libraries and the dataset
import pandas as pd
import matplotlib.pyplot as plt
import seaborn as sns
```

```
data = pd.read_csv("Data.csv")
```

We can print basic information about the dataset, like the shape of it, the column names, and even the first few rows to get an idea on what the data looks like. Sometimes, not all of the data would be useful for building a model and thus can be removed. However, for the present example such is not the case, as all columns are relevant to our study.

```
print(data.shape)
print(data.columns)
print(data.head(5))
```
Output:
```
(1001, 8)
Index(['Name', 'P1 _Convergence Length', 'P2 Divergence
Length', 'InletDiameter', 'P4_Throttle_Diameter',
'P5_ExitDiameter', 'Pressure_Inlet', 'P6_Velocity'],
      dtype='object')

Name  P1 _Convergence Length  P2 Divergence Length
P3_InletDiameter \
0  DP 0  15  15  12
1  DP 1  28  30  21
2  DP 2  19  17  13
3  DP 3  20  24  19
4  DP 4  13  14  11

   P4_Throttle_Diameter  P5_ExitDiameter  P6_Pressure_Inlet
P7_Velocity
0  3  15  1378950 119.130700
1  1  9   474973  25.174023
2  3  20  1285097 97.235954
3  1  23  382353  28.955469
4  3  13  822234  130.119600
```

There are plenty of other ways to do this, but for DataFrames you can use the 'describe' function to get some more useful information about the dataset with only one line. By the way, DataFrames are primarily provided by the Pandas library, and are fundamentally a two-dimensional structure for storing data. They are a convenient structure for storing datasets in machine learning and facilitate diverse preprocessing tasks (e.g. handling missing values, normalizing the data, etc.). This makes morphing the dataset relatively easy, which could be things like adding new rows, columns, or other manipulations. It would be worth your time to garner a basic familiarization with them, as they often show up in machine learning projects.

```
data.describe().apply(lambda s: s.apply('{0:.5f}'.format))
```

Output:

	P1	P2	P3	P4	P5	P6	P7
Count	1001.00	1001.00	1001.00	1001.00	1001.00	1001.00	1001.00
Mean	19.82	20.17	16.36	2.98	16.41	748911.00	117.71
STD	6.16	6.05	5.16	1.42	5.09	366097.00	64.22
Min	10.00	10.00	8.00	1.00	8.00	138829.00	24.24
25%	15.00	15.00	12.00	2.00	12.00	424986.00	65.79
50%	20.00	20.00	16.00	3.00	16.00	761254.00	111.13
75%	25.00	25.00	21.00	4.00	21.00	1052411.00	158.18
Max	30.00	30.00	25.00	5.00	25.00	1378950.00	375.67

Next, we will check for missing values in the dataset. There are various ways to check for missing data, but as an example here we can do a quick check for missing values and print the results in the output window. Since our dataset is too large to carefully inspect each row, we can rely on True/False statements to report if any missing values are present. 'Tricks' like this are things you will commonly use to save time in ML projects.

```
# Check for missing values (returns DataFrame whereby True
entries equates to missing data)
missing_values = data.isnull()
# Count the number of missing values in each column
(Returns a Series with the count of missing values in each
column)
missing_counts = missing_values.sum()
# Check if any missing values exist in the DataFrame
(returns True if any missing value is found)
```

```
has_missing_values = missing_values.any().any()

# Print the results
# print(missing_values) #Uncomment to see array
print(missing_counts)
print(has_missing_values)
```

Output:
```
Name                     0
P1 _Convergence Length   0
P2 Divergence Length     0
P3_InletDiameter         0
P4_Throttle_Diameter     0
P5_ExitDiameter          0
P6_Pressure_Inlet        0
P7_Velocity              0
dtype: int64
False
```

We want to plot distributions of our variables to better understand the shape, peaks, skew, and spread of the data. Among other things, this information can signal us to re-shape the data with transforms, remove outlier data or data outside specific ranges, and realize relationships between different individual variables. Creating these histograms and looking at their shape is a typical step in EDA, and noticing which transformations to apply has given me free jumps in accuracy of 10% for certain past projects.

```
# Set the size of the figure
fig, axes = plt.subplots(1, 7, figsize=(14, 3))
data.hist(ax=axes) # Create histograms for all columns
plt.tight_layout() # Adjust spacing between subplots
```

Output:

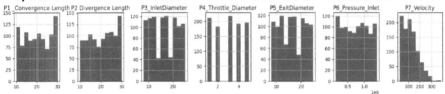

Upon plotting the raw distributions, we can notice that the ranges for each variable are different and, especially in the case of the inlet pressure, orders of magnitude different in scale from one another. Normalization is the process of scaling each variable by their minimum and maximum values so that all values are between zero and one for each variable. It is not always required to normalize the data but depending on the machine learning model being used it can be very useful. As one example, neural networks (and other models) use a gradient descent-based optimization algorithm (to minimize the model loss) which has better convergence when working with normalized data. Since gradient descent uses the feature value in the equation for determining the step size, if the input variables are not normalized to a uniform (or atleast similar) range, then the step sizes will be different for each feature. This can be quite difficult for convergence during training!

Another popular approach to transform your data is standardization, which uses the mean and standard deviation for each input variable so that the mean of each becomes zero and the distribution has a unit standard deviation. There is no hard rule for when to normalize or standardize, but personally whenever the data is mostly Gaussian (and as a result symmetric about the mean) I tend to opt for standardization. Of course, a safe approach is to fit your model with raw, normalized, and standardized data and then compare.

Lastly, there are other transformations (e.g. logarithmic) that are useful in scenarios where your data is skewed, has exponential relationships between variables, or needs compression because it scales such a large range. Pictured below are several of the aforementioned transformations for the velocity variable. In this case, the logarithmic transform is very useful to restore more of a normal distribution, which will likely improve the resulting model accuracy. Transforms such as this are helpful to improve the model accuracy when much of the data is clustered tightly together and unevenly.

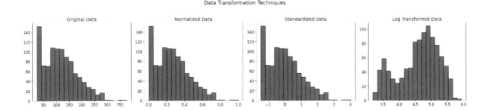

Figure 13: (From left-to-right) original data altered with several transforms: normalization, standardization, logarithmic transform.

Another popular way to visualize the data distribution is the box plot. In a typical box plot, the vertical axis represents the values of the variable being analyzed. The box itself shows the interquartile range (IQR), which spans from the first quartile (Q1) to the third quartile (Q3). The line within the box represents the median. The "whiskers" extend from the box to indicate the range of non-outlier values. Any individual data points falling outside the whiskers are considered outliers and are usually plotted separately as individual points.

```python
# Skip the last two columns
data_subset = data.iloc[:, 7]

# Create a box plot with individual data points
fig, ax = plt.subplots()
sns.boxplot(data=data_subset, ax=ax)
sns.stripplot(data=data_subset, color='black', size=4,
jitter=True, ax=ax)
plt.ylabel('Velocity')
plt.show()
```

Output:

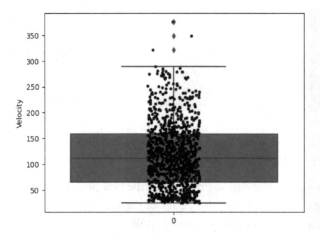

For a better understanding of our outliers, there are a few basic tools to reach for during EDA. For this dataset, which is rather small, it makes sense we can simply look at the histograms. For a quantitative measure to identify outliers in a dataset, which compliments the visual method of looking at plots and be useful in large datasets where visual inspection can be undesirable, we can use the IQR scoring technique as a guideline. This method is based on the IQR, which is a measure of statistical dispersion, representing the range between the first quartile (Q1) and the third quartile (Q3) of a dataset. By the way, we use quartiles to divide a dataset into four equal parts, representing points that split the data into quarters.

In this code, we can focus on applying this approach to the velocity column in particular, since it does look like higher velocity values are less common in the dataset. Once we carry through a few operations to calculate the quantiles, the IQR, and the lower and higher bounds, we can identify specific samples in the dataset which exceed the higher bounds (there are none below the lower bound in this problem FYI). These are cases 101, 157, and 830. It is reasonable to expect that removing some of these few datapoints can improve model accuracy.

```
Q1 = data.quantile(0.25)
Q3 = data.quantile(0.75)
IQR = Q3 - Q1

# Calculate lower and upper boundaries
lower_bound = Q1 - (1.5 * IQR)
upper_bound = Q3 + (1.5 * IQR)
```

```
# Let's focus on the velocity column
lower_bound_V = lower_bound[6]
upper_bound_V = upper_bound[6]

# Identify outliers
outliers = data[(data['P7_Velocity'].values <
lower_bound_V) | (data['P7_Velocity'].values >
upper_bound_V)]
outliers = outliers.iloc[:, [0, -1]] #remove all columns
other than P6_Velocity and Name
print(outliers)
```

Output:
```
        Name  P7_Velocity
101  DP 101     321.67303
157  DP 157     349.31204
830  DP 830     375.66592
```

Next during EDA, we will identify any existing correlation between different variables in our dataset (i.e. DataFrame). The '.corr()' function is a method with Pandas that calculates the pairwise correlation (resulting in a coefficient) between two columns (variables) in your dataset. The coefficient sign and number quantify the strength and direction of their linear relationship. The resulting matrix will display the Pearson correlation coefficients between the selected columns/variables. Values closer to 1 indicate a strong positive correlation, values closer to -1 indicate a strong negative correlation, and values close to 0 indicate a weak or no correlation between the variables.

This information can help in understanding dependencies, identifying patterns, and transform the data accordingly to produce better machine learning models. As one example, the correlation coefficients can assist in feature selection by identifying the most relevant variables for prediction. Variables with high correlation to the output variable(s) may be strong predictors and can be prioritized for model building, while variables with low correlation may be less useful and can be excluded to simplify the model.

High correlations between input variables can lead to issues in some machine learning models and is very important to notice. By using the correlation matrix to identify variables that are highly correlated we can address the multicollinearity (usually removing variables or introducing regularization).

```python
# Calculate the Pearson correlation
correlation_matrix = data2.corr()


plt.figure(figsize=(8, 6))
sns.heatmap(correlation_matrix, annot=True, linewidths=0.5)
plt.title('Correlation Matrix Heatmap')
plt.show()
```

Output:

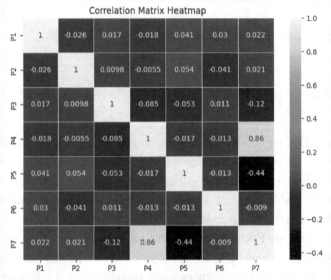

Figure 14: Pearson coefficient correlation coefficient. The parameter names have been shortened to save space.

After noticing more significant correlations in the velocity from P4 (throttle diameter) and P5 (exit diameter), which we do know from domain knowledge in fluid mechanics but have shown we could identify with purely typical data science analysis like Pearson correlation, we can make scatter plots for each variable versus the velocity. In the code output, which is a scatter plot, we can recognize a familiar relationship for the decay of velocity with increasing throat diameter for this flow regime. Having made this observation, we can achieve better machine learning predictions by starting with the relationship of these two variables evident in the figure.

```
fig, ax = plt.subplots(1, 7, figsize=(24,6))

for i,col in enumerate(data.columns[1:]):
    #random_df.plot(kind='scatter', x=col, y='MEDV',
ax=ax[i])
    sns.regplot(x=data[col], y=data["P7_Velocity"],
ax=ax[i])

fig.suptitle('My Scatter Plots')
fig.tight_layout()
fig.subplots_adjust(top=0.95)

plt.show()
plt.clf()
plt.close()
```

Output:

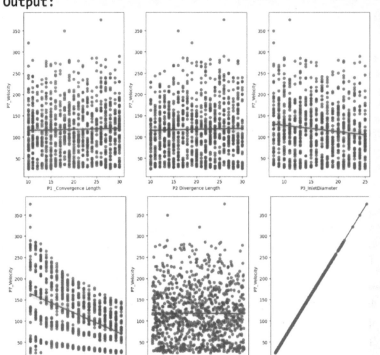

At this point, to cover more selected topics within the data exploration and cleaning stage, let's conclude the example on the converging-diverging nozzle example and shift to another case study: flow over an isothermal flat plate. Specifically, let's recall correlations for the friction coefficient for separate instances of both laminar and turbulent flow. We will use these textbook correlations [8] as a means for generating data, since we have a good understanding of them, for which we will analyze to illustrate more concepts in data exploration and cleaning.

Equation 1: Definitions for average Nusselt number and friction coefficient (for an isothermal plate with laminar flow)

$$Nu = \frac{hL}{k} = 0.664\ Re_L^{1/2}Pr^{1/3} \quad \& \quad C_f = \frac{1.328}{Re_L^{1/2}} \quad for \quad Pr \geq 0.6$$

Equation 2: Definitions for average Nusselt number and friction coefficient (for an isothermal plate with turbulent flow)

$$Nu = \frac{hL}{k} = 0.037\ Re_x^{\frac{4}{5}}Pr^{\frac{1}{3}}$$

$$\& \quad C_f = \frac{0.074}{Re_L^{\frac{1}{5}}} \quad for \quad 0.6 \leq Pr \leq 60 \ \& \ 5 \cdot 10^5 \leq Re_L \leq 10^7$$

Equation 3: The length (x_{cr}) over which the flow is laminar, For a critical Reynolds number of 5x10^5

$$Re_{cr} = 5 \cdot 10^5 = \frac{V_\infty x_{cr}}{\upsilon}$$

We will process this data with a t-SNE plot (t-distributed stochastic neighbor embedding). t-SNE is a dimensionality reduction technique that can be used for visualizing high-dimensional data in a lower-dimensional space that's easier to draw conclusions on (especially if we have more than three dimensions/variables to visualize). It's useful because it can help in revealing hidden structures, clusters, and patterns in your dataset. At this stage of data exploration and cleaning, t-SNE is just one of several tools with this capability. Why do we care about identifying patterns? Because if we can identify a pattern between in our dataset, for example a pattern between a selection of our inputs and the output, we can cast (manipulate) our input features an ideal way to reduce the data required, the inputs used, and the error in our final model, for example. After all, when we do exercises like AutoML, we are hoping that the technique will automatically identify these relationships in the data, which we can avoid if uncover these relationships ourselves. We demonstrate exactly this in the following two figures.

In this example, we have generated two classes of data from the correlation: laminar and turbulent (we ignore transitional). Usually, one would be dealing with more dimensions than two when generating a t-SNE plot, so this result will be simple to analyze. With the below code block, we create the t-SNE plot and can see the resulting red and blue data clusters are completely segregated into different sections inside the plot area.

Data which is segregated in the resulting t-SNE plot indicates that the inputs can clearly be mapped to the separate outputs. Usually, this is a good thing and indicates that the choice of input variables is well selected such that the output values can be accurately predicted from the inputs. Conversely, if many different colored dots are all overlapping without forming clear clusters, then there is not ample evidence to claim that those input variables can accurately be fit by a machine learning model to the outputs (which means it might be hard to build an accurate ML model with those inputs).

```python
import pandas as pd
from sklearn.cluster import KMeans


# Load the regression dataset
import pandas as pd
import numpy as np
import matplotlib.pyplot as plt
import seaborn as sns
from sklearn.manifold import TSNE
from sklearn.preprocessing import StandardScaler


# Separate the classification column
classification = data['Regime']


# Select the continuous columns and convert them to
numerical data type
continuous_columns = ['Vinf', 'L', 'Re', 'Cf']
continuous_data = data[continuous_columns].astype(float)
```

```
# Scale the continuous data
scaler = StandardScaler()
scaled_data = scaler.fit_transform(continuous_data)

# Perform t-SNE on the scaled data
tsne = TSNE(n_components=2, random_state=42, perplexity=2,
learning_rate=200, n_iter=5000)
tsne_result = tsne.fit_transform(scaled_data)

# Create a new DataFrame with the t-SNE results and the
classification column
tsne_df = pd.DataFrame(tsne_result, columns=['t-SNE
Component 1', 't-SNE Component 2'])
tsne_df['Regime'] = classification

# Plot the t-SNE scatter plot with different colors for
each group
plt.figure(figsize=(8, 6))
sns.scatterplot(data=tsne_df, x='t-SNE Component 1', y='t-
SNE Component 2', hue='Regime', palette='Set1')
plt.title('Perplexity = 2')
plt.show()
```

Output:

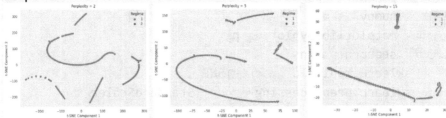

Figure 15: t-SNE plots, which vary perplexity from 2 to 5 to 15, for the laminar (regime=1) and turbulent (regime=2) data samples from equation.

Often, once t-SNE has been done to transform the dataset into a lower dimensional space, the data is given to clustering algorithms. However, since we don't have too many dimensions in this case, we can work from the original dataset. Clustering can be done to identify clusters or groups of similar data points, which is helpful to find natural groupings or patterns within the data based on their similarities. In this case, we are going to group the samples into two clusters with the k-means clustering algorithm (very popular) and then compare the results to our physical knowledge. Since we know that friction will depend on the flow regime, meaning if it is laminar or turbulent, the clusters formed should provide one for the laminar samples and then another for the turbulent samples. What's great about this example is that we know the phenomenon which differentiates the samples, and we can even use a critical Reynolds number of 500k to label each sample as laminar or turbulent. That way we can compare the resulting clusters to see which samples were incorrectly grouped (which turbulent cases were grouped with laminar ones, and vice versa). We are using our knowledge of fluid mechanics to test out how well these clustering methods can 'blindly' (or rather uninformed) produce realistic segregations of the data.

```python
import pandas as pd
from sklearn.cluster import KMeans

# Load the regression dataset
data = pd.read_csv('Data.csv')

# Select the features for clustering
features = ['Vinf', 'Re', 'Cf', 'L', 'v']
X = data[features]

# Perform clustering using K-means algorithm
k = 4   # number of clusters
kmeans = KMeans(n_clusters=k, random_state=42)
labels = kmeans.fit_predict(X)

# Add the cluster labels to the dataset
data['Cluster'] = labels

# Print the cluster assignments
print(data[['Vinf', 'Re', 'Cf', 'L', 'v', 'Cluster']])
```

Output:

	Re	Cf	L	v	Cluster
0	20661	0.009239	5.00	0.000242	2
1	24793	0.008434	6.00	0.000242	2

2	28926	0.007808	7.00	0.000242	2
3	33058	0.007304	8.00	0.000242	2
4	37190	0.006886	9.00	0.000242	2
..	...	...	...	...	...
202	1279339	0.004445	2.15	0.000242	1
203	1318182	0.004418	2.20	0.000242	1
204	1357438	0.004392	2.25	0.000242	1
205	1397107	0.004367	2.30	0.000242	1
206	1437190	0.004342	2.35	0.000242	1

[207 rows x 5 columns]

Our new DataFrame will include an added column with the cluster label, which we can compare to the first column 'Regime'. What's interesting is that we know our friction coefficient result is related to our input variables via the textbook correlation from our domain knowledge on fluid mechanics. When using machine learning models, it is their purpose to uncover an accurate mapping between the friction coefficient and the input variables on their own, which is a different means to arrive at an understanding similar to the correlation equation. As such, we know that deciding to include Reynolds number directly, or just the individual variables inside the Reynolds number equation, could vary the difficulty in the clustering algorithm to accurately segregate the laminar and turbulent cases. Indeed, there are more incorrect cases when not directly including the Reynolds number, which is summarized in Figure 16.

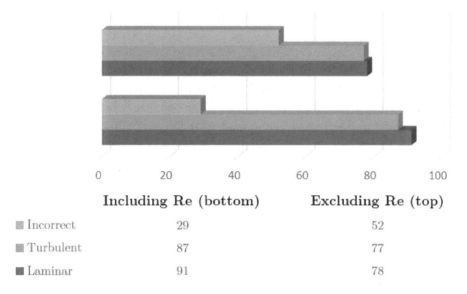

	Including Re (bottom)	Excluding Re (top)
▣ Incorrect	29	52
▣ Turbulent	87	77
▣ Laminar	91	78

Figure 16: Summarizing the flow regime predictions via two clustering models, which differ based on including Reynolds number in the input variables (or not but the individual variables inside the definition).

What's illustrative about this example is that if we do not know the underlying relationship between the inputs and outputs, we would have to go about attempting to find it in a blind way. While we don't necessary seek to determine the full relationships during the data exploration phase, as that's basically the purpose for model training and tuning, it's clear from this example that it would be beneficial to try different variations of our inputs and move forward to model training with the permutation that appears most suitable. This is essentially an organic realization of one of the goals in what's called feature engineering.

Furthermore, this example was also selected to provide a cautionary suggestion: be observant for situations where the input variables are correlated to one another (as in velocity and Reynolds number in this example). This often goes by the term collinearity, or multicollinearity, and is a well-known challenge in the machine learning community. If you observe such in your project, you'll want to make modifications to the dataset to avoid correlation between your inputs. Not only is it redundant in terms of information, which can add unnecessary complexity to the modeling, but it can also yield overemphasis on the correlated features and instability. Further, it impairs the user's ability to interpret the model behavior. Some popular approaches to start with to reach a resolution are Linear Discriminant Analysis (especially when dealing with classification tasks), techniques like Ridge Regression and LASSO (especially for regression problems), or dimensionality reduction techniques like Principal Component Analysis.

In a simple sense, the goal of feature engineering is to provide insights and methods on how we can improve and manipulate our collected dataset to be more ideally posed for our problem at hand. The outcome could be a more accurate model, a model which can more quickly compute with lesser hardware resource or train faster, or a more interpretable model. Feature engineering can help us avoid large, black-box, complicated models and instead generate critically engineering features that are performant with simple models. Ensure the concept of feature engineering is clear as a means to create new features and also transform the existing data.

Before going forward, one point of clarification needs to be made: "features" and "inputs" are often used interchangeably, which is also fair in the context of this book. To be thorough, features are technically a specific type of input data, such as manipulations done to the input data, whereas input refer to the broader set of data or variables used in the machine learning process.

Since there are a wide variety of types of machine learning (regression and classification, supervised and unsupervised, etc.) and a large diversity of data types (numerical, categorical, etc.), the topic of feature engineering is understandably vast. For the CAE domain, naturally some principles and methods within feature engineering will be more commonly used. Before moving to such concepts, we will briefly list a few mainstream methods across many different domains. After all, when we combine data from different sources, like simulation data, user data, and testing data, it is possible that some of these techniques can become quite relevant and helpful.

- **Handling missing values.** Sometimes entries in the dataset will be incomplete, whereby you may have missing samples or missing values within an individual sample. The treatment of such could be as simple as deleting the entry or inferring missing values to salvage the data sample. For the latter, the idea is to use available data to infer reasonable values for those missing. Options could be the mean/median/mode of the feature column, or even using a regression model to predict missing values based on other features.

- **One-hot encoding.** Conversion of categorical variables (e.g. names) into numerical values. For example, if data is collected in a wind tunnel by three different researchers then instead of using their names there could be encoded values of 1, 2, and 3 for each respective person.

- **Binning**. Converting continuous variables into discrete groups, which can help handle extreme values and/or non-linear relationships by segregation into separate bins. For this plate example, different bin designations could be assigned based on Reynolds number (0-100k: 1, 101k-500k: 2, etc.)

Let's review a process of steps for a feature engineering pipeline which could be considered mainstream and widely used. To demonstrate this, and subsequent steps of feature selection (removal), we can use the California housing price dataset from scikit-learn.

```
from sklearn.datasets import fetch_california_housing
import pandas as pd
import numpy as np

#intake data
data = fetch_california_housing()
X = data["data"]
col_names = data["feature_names"]
y = data["target"]

df = pd.DataFrame(X, columns=col_names)
```

First, just for the sake of demonstration, we can make a few specific manipulations to existing features in the dataset and use them to create new columns. Our manipulations will include basic operations on individual columns, as well as operations using multiple columns at once. We can then use the Pearson coefficient matrix to observe how correlated the new features, but since these manufactured columns will probably need to be remove because we derived them from existing columns. Again, since collinearity between features is bad for the accuracy of the resulting model! Such columns which are related carry similar information, and their existing relationship can negatively impact model performance and interpretability.

```
df['MedInc_Log'] = np.log(df['MedInc'])
df['MedInc_Exp'] = np.exp(df['MedInc'])
```

```
df['HouseAge_Squared'] = df['HouseAge'] ** 2
df['Interaction'] = df['MedInc'] * df['AveRooms']

from sklearn.preprocessing import MinMaxScaler
scaler = MinMaxScaler()
df['Population_Normalized'] =
scaler.fit_transform(df['Population'].values.reshape(-1,
1))

# Columns to keep in the correlation matrix
columns_to_keep = ['MedInc', 'HouseAge', 'AveRooms',
'Population', 'MedInc_Log', 'MedInc_Exp',
'HouseAge_Squared', 'Interaction']

# Compute correlation matrix
correlation = df[columns_to_keep].corr()

correlation
```

Output:

	MedInc	HouseAge	AveRooms	Population	MedInc_Log
MedInc	1.000000	-0.119034	0.326895	0.004834	0.938688
HouseAge	-0.119034	1.000000	-0.153277	-0.296244	-0.139192
AveRooms	0.326895	-0.153277	1.000000	-0.072213	0.314303
Population	0.004834	-0.296244	-0.072213	1.000000	0.032068
MedInc_Log	0.938688	-0.139192	0.314303	0.032068	1.000000
MedInc_Exp	0.349009	0.039339	0.084591	-0.033654	0.189294
HouseAge_Squared	-0.101859	0.973022	-0.138689	-0.272226	-0.123800
Interaction	0.869603	-0.151144	0.689763	-0.030640	0.776515

MedInc_Exp	HouseAge_Squared	Interaction
0.349009	-0.101859	0.869603
0.039339	0.973022	-0.151144
0.084591	-0.138689	0.689763
-0.033654	-0.272226	-0.030640
0.189294	-0.123800	0.776515
1.000000	0.045589	0.365781
0.045589	1.000000	-0.130000
0.365781	-0.130000	1.000000

Next, synthetic features can be created to increase the total number of features considered as inputs to the machine learning training process. Ultimately, creating synthetic features can be a powerful step to reduce overfitting by using more ideal features than the original set. *PolynomialFeatures* is a popular function in scikit-learn to produce more features, which takes all possible permutations of your existing features by combining them together sequentially up to a specified power. Note that in this process one would rarely use all the new features generated, which could be hundreds or even thousands of features, but rather we normally down-select those features to a small set which are most effective.

In the example below, we will create polynomial features up to the third power and then fit a simple linear regression model. By fitting a model, we can then observe the hierarchy of which features were most significant to the resulting model. The features will be shown after a sorting is done, from high to low feature importance, and the top ten values will be printed on the plot so one can get a sense of scale when looking at the log-based horizontal axis of the plot. When using different machine learning model libraries, there will often be unique and specific functionality for each that can report the feature importance. It's a useful thing to do in order to gather insight on the behavior of your model.

```python
import numpy as np
import pandas as pd
from sklearn.datasets import fetch_california_housing
from sklearn.preprocessing import PolynomialFeatures
from sklearn.linear_model import LinearRegression
import matplotlib.pyplot as plt
```

```python
# Load the California housing dataset
data = fetch_california_housing(as_frame=True)
X = data.data
y = data.target

# Apply PolynomialFeatures
poly = PolynomialFeatures(degree=3, include_bias=False)
X_poly = poly.fit_transform(X)

# Get feature names
feature_names = [f'feature_{i}' for i in
range(X_poly.shape[1])]

# Create a DataFrame with polynomial features
X_poly_df = pd.DataFrame(X_poly, columns=feature_names)

# Fit a linear regression model
model = LinearRegression()
model.fit(X_poly, y)

# Get feature importances
importances = np.abs(model.coef_)

# Sort feature importances in descending order
sorted_indices = np.argsort(importances)[::-1]
sorted_importances = importances[sorted_indices]

# Plot feature importances
plt.figure(figsize=(10, 6))
plt.bar(range(len(sorted_importances)), sorted_importances)
plt.xticks(np.arange(0, len(sorted_importances), 5),
np.arange(1, len(sorted_importances) + 1, 5),
rotation='vertical')
plt.xlabel('Feature Number')
plt.ylabel('Importance')
plt.title('Feature Importance')

# Print top 10 features and importances
```

```python
top_features = range(1, 11)
top_importances = sorted_importances[:10]

# Calculate the maximum importance value
max_importance = np.max(sorted_importances)

# Position for the stacked labels
label_x = len(sorted_importances) + 1
label_y = max_importance + 0.03

# Stacked labels
for i, (feature, importance) in enumerate(zip(top_features,
top_importances)):
    plt.text(
        label_x, label_y - i * (max_importance / 10),
        f'Feature {feature}: {importance:.4f}',
        ha='right'
    )

plt.xscale('log')
plt.tight_layout()
plt.show()
```

Output:

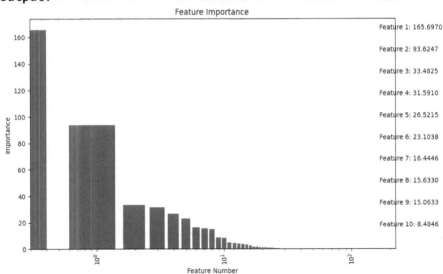

Thankfully, we don't have to fully rely on the feature importance's alone, but we can also do other experiments to better ascertain which features to select. One such technique is to look at those features which have the highest variance from the full set created. Features with low variance may not provide much discriminatory information and could be removed to simplify the model. You can set a variance threshold and select features above that threshold, or simply take the highest ones.

Confused on why variance is important? Let's give an example- I like to give introductions to new concepts on health/body related examples because I feel they are intuitive. Let's say we want to build a machine learning model that can identify Olympic athlete potential in youth athletic populations. If we wanted to construct a dataset, we could pick one feature to describe the athlete's hair. Somewhat silly, sure, but it is not a crazy idea; early facial hair/balding in youth males can be related to testosterone production (DHT), which both have an influence on muscle building and power potential. In that sense, maybe there is some value for including one feature on this, and the variance might be non-trivial for such a feature (and therefore worth including). However, how much benefit and unique information is added to the resulting model if we include five features more on the athlete's hair? I imagine very little, and as such the subsequent variances for the five features would be very low.

Another option is Recursive Feature Elimination, which is an iterative technique of training the model multiple times, removing the least important features in each iteration. The process continues until the desired number of features is reached or the model's performance plateaus.

Building from the previous code, the plot shows features sorted into descending variance. The drop-off is less steep than for the feature importance's, which can pose a challenge in only selecting a small number of features.

```
# Calculate the variance for each feature
variances = np.var(X_poly, axis=0)

# Sort variances in descending order
sorted_indices = np.argsort(variances)[::-1]
sorted_variances = variances[sorted_indices]
```

```
# Plot variances
plt.figure(figsize=(10, 6))
plt.bar(range(len(sorted_variances)), sorted_variances)
plt.xticks(np.arange(len(sorted_variances)), feature_names,
rotation='vertical')
plt.xlabel('Feature')
plt.ylabel('Variance')
plt.title('Feature Variance')

plt.xscale('log')  # Set x-axis to log scale
plt.yscale('log')  # Set y-axis to log scale

# Calculate the 99% value of the top feature variance
top_feature_variance = sorted_variances[0]
threshold_99 = np.percentile(sorted_variances, 99)

# Count features with less variance than the 99% value
count_less_than_99 = np.sum(sorted_variances <
threshold_99)

# Plot the 99% value and count inside the plot area
plt.text(len(sorted_variances) * 0.7, threshold_99 * 0.9,
f'99% value of max variance:\n{threshold_99:.2e}',
ha='right', va='bottom')
plt.text(len(sorted_variances) * 0.7, threshold_99 * 0.8,
f'Features < 99%: {count_less_than_99}', ha='right',
va='top')

plt.tight_layout()
plt.show()

# The code below is for the table summary below

# Print the top 10 features by importance
print("Top 10 Features by Importance:")
for i in sorted_importance_indices[:10]:
    print(f"Feature {i}: Importance =
{sorted_importances[i]}")
```

```
# Print the top 10 features by variance
print("\nTop 10 Features by Variance:")
for i in sorted_variance_indices[:10]:
    print(f"Feature {i}: Variance = {sorted_variances[i]}")
```

Output:

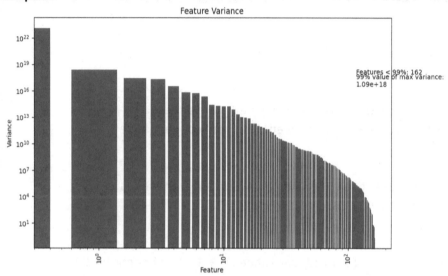

Finally, we can output values for the top ten features from both exercises by sorting by importance and variance. For the housing price dataset, we perform several exercises of training the model based on different feature sets crafted: the original features, all features created from the exercise of creating polynomial features up to the third power, combinations of the top ten features from importance and variance, and finally features autonomously selected when applying a Greedy algorithm. Without any careful fine tuning of the selection of features, and while also using a very simple linear regression model, the only case which produced a mean square error lower than the original set was that which used the Greedy algorithm for feature selection. Unlike this simple demonstration, more serious feature engineering efforts often yield significantly improved model predictions as a result (of course not with linear models).

Top 10 Features by Importance:	Top 10 Features by Variance:
Feature 6: Importance = 16.44	Feature 144: Variance= 3213
Feature 7: Importance = 15.63	Feature 147: Variance = 1588
Feature 2: Importance = 33.48	Feature 145: Variance = 2463
Feature 29: Importance = 0.21	Feature 146: Variance = 1882
Feature 32: Importance = 0.19	Feature 98: Variance = 1.9E6
Feature 0: Importance = 165.7	Feature 148: Variance = 702.2
Feature 5: Importance = 23.10	Feature 119: Variance = 1.9E5
Feature 3: Importance = 31.59	Feature 70: Variance = 6.7E7
Feature 24: Importance = 0.75	Feature 153: Variance = 320
Feature 1: Importance = 93.62	Feature 134: Variance = 23834

MSE	Dataset
0.55	Original Features
19.97	All features
0.77	Top ten features by importance
1.26	Top ten features by variance
0.65	Top ten features by importance + variance
0.51	Greedy algorithm feature selection

Another popular approach for feature selection is to perform univariate analysis to score each feature against the target, which helps to identify features that are most informative in predicting the target variable. Some mainstream methods for univariate feature selection are Mutual Information (measures dependency between two variables), ANOVA F-test (Analysis of Variance), and the Chi-square test (for categorical target variables and features).

Up until now, we have talked about ways to ideally craft a set of input features to be optimal for fitting a machine learning study. This includes transforming current features, creating synthetic ones, and also down-selecting a critical group of features based on things like feature importance. Next, we can talk about compression techniques, which alternatively will be yet another approach we can take to help our model fitting experience be more pleasant (more accurate, lower memory, faster, etc.). Compression techniques, such as Principal Component Analysis (PCA) or Variational Autoencoders (VAE) are popular due to their ability to reduce the dimensionality of data while retaining most of its important information. PCA is a linear dimensionality reduction using Singular Value Decomposition of the data to project it to a lower dimensional space [9]. You may have heard of it before in your simulation career, as it's a popular technique to break down the physics simulation results by the highest energy modes to analyze the main contributions towards the final result. Such benefits are the convenience had when able to reduce the dimensionality of high-dimensional large datasets, filtering out noise that may exist in imperfect 'real world' datasets, helpful analysis when viewing in fewer dimensional space, less memory required, and avoiding collinearity, to name a few.

Let's use the famous wine dataset, which contains the results of chemical analysis of various wines with respect to their resulting quality, for demonstrating a basic application and analysis of PCA (with some basic comparison to a VAE made from the same data). Using Pandas, we can import this dataset and do some basic cleaning (e.g. removing NaN values) and transformations (standardization). You'll notice the output provided below from a basic 'data.head(5)' command. With sklearn.decomposition, we can use 'PCA(n_components=3)' to create and print a new DataFrame.

Alcohol	Malic.acid	Ash	Acl	Mg	Phenols	Flavanoids	Nonflavanoid.phenols	Proanth	Color.int	Hue	OD	Proline
14.23	1.71	2.43	15.6	127	2.80	3.06	0.28	2.29	5.64	1.04	3.92	1065
13.20	1.78	2.14	11.2	100	2.65	2.76	0.26	1.28	4.38	1.05	3.40	1050
13.16	2.36	2.67	18.6	101	2.80	3.24	0.30	2.81	5.68	1.03	3.17	1185
14.37	1.95	2.50	16.8	113	3.85	3.49	0.24	2.18	7.80	0.86	3.45	1480
13.24	2.59	2.87	21.0	118	2.80	2.69	0.39	1.82	4.32	1.04	2.93	735

After the data is loaded in, it takes just under ten lines of code to complete the PCA transformation with the final step being 'pca.fit_transform(x_pca)'. We can store the principal components as separate columns in a new DataFrame and then set it aside to show alongside the VAE results (which is much more work to setup, at about ten times the lines of code as the PCA). For the sake of illustration, the outputs are shown below for the first few rows in the (separate) DataFrames for both the PCA and VAE. A convenience of this approach is reducing the number of independent variables to a more manageable size to sample. It's not uncommon to have dozens of geometric parameters to describe your simulation, which would equate to having dozens of columns (one for each variable, whereby the rows are unique samples). With these types of examples, you can see the compression to fewer columns with an equal number of rows before and after the transformation (the number of simulations generated).

PCA Output:
```
        principal component 1     principal component 2
```

	principal	component 3	
0	3.316750	-1.443462	-0.165739
1	2.209467	0.333392	-2.026458
2	2.516741	-1.031153	0.982821
3	3.757066	-2.756373	-0.176191
4	1.008908	-0.869830	2.026688

VAE Output:

	latent_dim_1	latent_dim_2	latent_dim_3
0	1.653812	0.222876	0.445429
1	1.344676	-0.197746	-0.786440
2	1.465320	0.161750	0.409438
3	2.092077	0.388037	0.613449
4	0.633846	0.602323	1.213200

The inherent structure of the dataset is shown for the respective techniques in both plots in the Figure 17 below. The scatter plots generated from the compressed representations offer a concise visualization of the complex relationships among the samples. A meaningful characteristic in the plots to look for is whenever datapoints (of the same color) are segregated from the other points. These clusters of separated points can represent salient patterns, similarities, or differences among the data points (similar to t-SNE). The segregation indicates that the algorithm has successfully captured and highlighted these underlying structures by reducing the data's dimensionality. In short, it means there is a pattern in the data coherent with those data samples. As such, it would be interesting to grab the specific samples in the VAE plot that form the yellow line-like pattern that moves away from the other clusters, as well as additionally those that appear planar in the purple series (principal component), and try to associate those points to any trends in the dataset (e.g. high values, or low values, or those clustered together in the original feature-space, etc.).

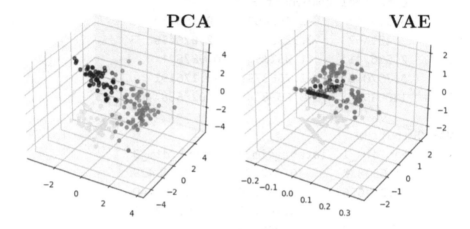

Figure 17: Compressed representations of the data through principle component analysis and variational autoencoders.

Without going into too much detail, since the present work is meant as an overview of many topics, we can at least cover a few concepts for how to evaluate such models before progressing to the next subject (with the note that this analysis can indeed go much deeper). One consideration is to evaluate the effectiveness for the dimensionality reduction performed, with either approach, to determine how much of the essential information was preserved. The metrics one can do this with, for example to compare VAE and PCA results, are the reconstruction accuracy or the explained variance. Higher values of either indicate better preservation of information. In PCA, the principal components are chosen in descending order of their corresponding eigenvalues, which represent the variance of the data along those components. The higher the eigenvalue, the more significant the variance captured by the corresponding principal component. Similarly, in VAE, minimizing the loss encourages the VAE to capture and retain the most important features of the data, which, in turn, translates to maximizing the variance in the compressed space.

In the next step of the code, the explained variance is obtained from explained_variance_ratio_ attribute of the PCA model, and the reconstruction accuracy is calculated using the custom loss function customLoss() applied to the VAE model's reconstructions and the original data. While VAEs do not provide an equivalent measure of explained variance like PCA, you can still analyze the distribution of the latent variables. If the VAE's latent space is effective, it should have learned to represent the main modes of the data distribution, and the variance across different dimensions of the latent space could indirectly indicate the importance of each dimension in representing the data.

Output:
```
PCA Explained Variance: [0.36198848 0.19207492 0.11123632]

Variance of each dimension in the latent space:
Dimension 1: 0.8503484725952148
Dimension 2: 0.013263336382806301
Dimension 3: 1.069663405418396
```

As a last remark on feature engineering, I would like to tie in a thought-provoking idea from some concepts we have learned in fluid mechanics. Let's reference the beautiful image on the cover I created back when doing my doctoral dissertation [43] in Figure 18. The image shown is a result of running a CFD simulation on the famous benchmarking case for an array of cylinders in a cross-flow duct by Ames [44], and post-processing some machine learning features to plot in the mid-plane.

When picking our features, they should be rich and extensive enough to help each sample in our dataset be distinct in the feature space, while also enabling us to have a meaningful relationship between the inputs and outputs in the correlation we make with a machine learning model. Thus, it is attractive to consider several mathematical operations we could apply on the mean flow variables (tensors) that should theoretically provide us a vast set of all possible characteristics of the mean flow. In other words, we can take dense tensors that describe our mean flow-field and do 'some' operations on it to reveal innumerous characteristics about our flow, and then use those as individual inputs to a feature engineering exercise so we can see which variables are most helpful in constructing a machine learning model. In the words of my mentor in 2017 "wouldn't it be cool to throw all the parameters and terms we know from fluid mechanics at a machine learning model and let it engineer which derived quantities are actually the most salient?".

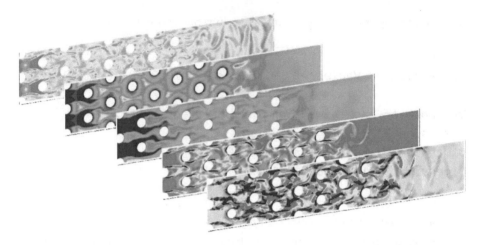

Figure 18: Plotting different scalars (right-to-left: trace of strain tensor to the third power, trace of the strain rate tensor to the second power, turbulent viscosity ratio, wall-based Reynolds number, ratio of turbulent time scale to mean strain time scale) on CFD results for flow through a duct with wall-mounted pin-fins [43].

So let's talk about these mathematical operations and the features we could create. We can use strain rate tensors and rotation rate tensors for each point (cell) in our computational domain for a calculation, as they are often terms used in eddy viscosity-based turbulence models and should be relevant enough to contain data which allows us to 'learn' a flow-field. These tensors are based on the velocity gradient tensor, which itself is a measure of how the velocity varies in space. We will use invariants of both of these tensors to reveal underlying characteristics of our flow that remain unchanged under coordinate transformations, as well as changes in observer's perspective (reference frame). While I am far from a mathematician, even I can appreciate the beauty in the physical significant of the invariants for strain rate tensor; the first invariant represents the rate of volumetric expansion/contraction of a fluid element, the second invariant relates to energy dissipation due to viscosity of a fluid element, and third provides some description of the three-dimensional deformation characteristics of the fluid element.

If we perform these operations on both tensors and do polynomial combinations of the different resulting products (for example) the third power, we can start to make a good list of new variables to use in a feature engineering study that should be rich with information. Further, we can also make our list longer by adding parameters that encapsulate our physical knowledge of fluid mechanics. These could be things like wall-based Reynolds number, the turbulent intensity, the eddy viscosity, pressure gradients, as well as others.

Inspired by several seminal works in the field of physics informed machine learning [43, 45, 46, 47], the image below shows a similar (small) list of features we could consider in the present discussion.

Variable	Definition
Turbulent Viscosity Ratio	μ_T / μ
Ratio: rotation rate to strain rate (Q-criterion)	$0.5 \cdot \left(\|\Omega\|^2 - \|S\|^2 \right)$
Wall-based Reynolds number	$d\sqrt{k} / \upsilon$
Pressure gradient along streamline	$\dfrac{\partial P}{\partial x_k}$
Ratio of time scales: turbulence and mean strain	k / ε
Ratio of convection to turbulence production	$U_i \dfrac{dk}{dx_i}$
Invariants with strain rate and rotation rate tensor [tr=trace]	$tr(S), tr(S^2), tr(S^3), tr(R^2), tr(R^2 S),$ $tr(R^2 S^2), tr(R^2 S R S^2)$

Figure 19: Sample derived parameters in fluid mechanics that can be used in feature engineering exercises [43, 48].

In conclusion, this is one example of how to aggregate variables that have relevant information into a feature engineering exercise to 'data mine' which of them could be the best to fit in a machine learning study, and possibly have a machine learning model create new features based on these variations (e.g. polynomial feature sets). It's quite a creative and exciting exercise, in my opinion, since it sits at the union of our application knowledge (fluid mechanics mathematics and our experience with dimensional parameters) and machine learning techniques, which could indeed yield new insights!

Next, let's shift the discussion to scenarios where the dataset available is suffering from being too small (and what actions we can take to make some improvements). Whether it feels like it is simply not enough samples to cover the relevant ranges of each parameter, or it is an imbalanced dataset which misleads the model predictions, or that you have limited samples that make dividing the dataset into subsets for validation tricky, it's a critical task to make insightful conclusions on how the dataset can limit the accuracy and generality of the trained model. Simulation and physical testing data is quite expensive to come by, and time is often too limited with industry work to generate generous amounts of data, so naturally you will likely have to deal with datasets of limited size at some point in your career. We will proceed to cover some examples with code snippets and then finish with a brief discussion on other considerations; there are way too many to review individually!

The dataset we will use will be purely for illustration purposes and not factually correct whatsoever. Why do this? Well, we can actually make this pseudo dataset with a large language model (e.g. ChatGPT4). I am highlighting this idea because it is a very attractive tool for creating sample datasets for illustrating different steps (for example pre-processing actions in a data science pipeline) or demonstrations that you may want to make in the future. Nothing with *actual* data, but often we can easily fulfill our tasks with pseudo data to illustrate a workflow, to make a test run for some code, or for other purposes. I thought it was a very clever resource and am happy to demonstrate it. Without conveying the full conversation with the chat bot, I essentially asked for a fake CFD dataset that is tabular, small in nature, and highly imbalanced. ChatGPT created a sample dataset for flow over a cylinder that had columns for velocity, Reynolds number, pressure, and flow_type (laminar or turbulent). With a few simple prompts we can request a few modifications, such as a certain number of rows, and then copy the data into a DataFrame into your coding environment. Voilà!

For this demonstration, let's start with the proposition of data augmentation; specifically, adding noise to the data in hopes that it would improve the accuracy of our model. This slightly perturbed data should, in theory if we are careful, help train the model to make more robust predictions and generalize better to unseen data for inference. We will create new samples by applying transformations to the original data, which in this case for numerical data can be small random noise. For other problems, for example with images, common modifications are adding random noise, rotations, scaling, warping, among others.

The add_noise_to_velocity function directly adds Gaussian noise to the "Velocity" column of the DataFrame. The noise_level parameter controls the standard deviation of the Gaussian noise, so you can adjust it to add more or less noise. In this simplified case, you can see the linear change in velocity throughout the population. In a more realistic case, there would be sharper gradients in trends for velocity throughout the population and the added noise would help prevent overfitting of such.

```
def add_noise_to_velocity(df, noise_level=0.01):
    noise = np.random.normal(0, noise_level,
df['Velocity'].shape)
    df['Velocity'] += noise
    return df


# Adding noise to the Velocity column
data_noisy = add_noise_to_velocity(data.copy(),
noise_level=10) # Adjust noise level as needed
```

Output:

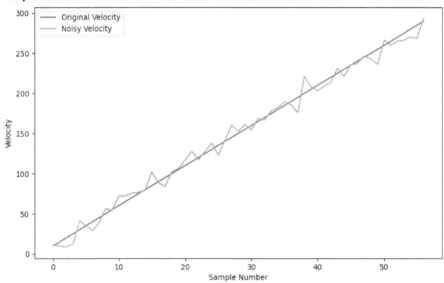

Next, we can discuss bootstrapping; another helpful resource in scenarios where you have a small or very limited dataset. Bootstrapping involves creating new datasets by sampling the initial one with replacement from the original dataset. This can help us when we want to estimate the distribution of a statistic and it can also be used for model validation as well. At a higher level, the process is: draw n number of samples from your dataset, then calculate your statistic of interest (e.g. standard deviation), then repeat this process (on the full and original dataset) many times (1,000 times is not uncommon) and analyze the resulting distribution of your samples from the 'bootstrap' process. It harvests the original dataset iteratively as the 'population' from which to draw samples and make inference. In the plot below we can see some of the core pieces in code snippets and then a plot to show the original data alongside that of a bootstrapped sample.

```
# Function to perform bootstrap sampling - The
bootstrap_sample function is used to create a new dataset
by sampling with replacement from the original dataset.
y = X['Velocity']
X_features = X.drop('Velocity', axis=1)

def bootstrap_sample(X, y):
    indices = np.random.choice(X.shape[0], size=X.shape[0],
replace=True)
    return X.iloc[indices], y.iloc[indices]

# Creating a bootstrap sample
X_bootstrapped, y_bootstrapped =
bootstrap_sample(X_features, y)
```

Output:

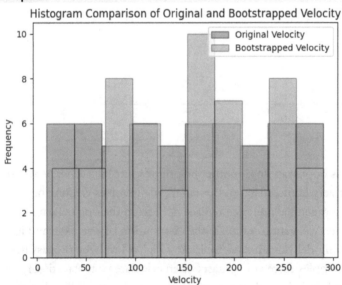

Histogram Comparison of Original and Bootstrapped Velocity

As briefly introduced above, one common challenge in small datasets is how to effectively split the full dataset for training, validation, and testing. If you only have for example 50 samples, it might feel painful to train on even less samples after breaking your dataset into partitions, and further depending on how you segregate the samples into each bucket it can have a large impact on the model training. After all, if we have plenty of samples then random selection should normally be fine, but if we have a very limited size dataset then the specific samples picked might be quite influential on the resulting accuracy of the model—what if we pick that one weird outlier point to evaluate out model accuracy! Often, train-test splits can leave too little data for either training or validation, leading to overfitting or validation metrics that aren't fully trustworthy.

Cross-validation can help in this scenario (as a resampling technique). Cross-validation splits the data into different training and validation sets, named folds (or into 'k' number of folds if you ever hear the term k-fold cross-validation), so that every sample can be used for validation exactly once. On the same ChatGPT fake dataset for flow over a cylinder, we can write a few lines of code to perform a 3-fold cross-validation on the dataset with a random forest model. From the code shared, the cv=3 argument in the cross_val_score function specifies that 3-fold cross-validation should be used. This means the dataset will be divided into 3 equal parts or "folds." Two of the folds will be used for training, and the remaining fold will be used for validation. This process is repeated 3 times, with each fold serving as the validation set once.

Note that in this case one variable is the flow regime, which has the value of 'turbulent' or 'laminar', which is not a numeric value as the other entries in the dataset. Therefore, we need to employ a one-hot encoding, which in this case is as simple as turbulent samples receiving a '0' value, and '1' for a laminar case. The beginning portion of the code below takes care of that.

```
# One-hot encode the "Flow_Type" column
encoder = OneHotEncoder(sparse=False,
handle_unknown='ignore')
encoded_features =
encoder.fit_transform(X_features[['Flow_Type']])
X_features_encoded =
pd.concat([X_features.drop(columns=['Flow_Type']),
```

```
pd.DataFrame(encoded_features,
columns=encoder.get_feature_names_out(['Flow_Type']),

 index=X_features.index)], axis=1)

# Create and train the model
model = RandomForestRegressor(random_state=42)
scores = cross_val_score(model, X_features_encoded, y,
cv=3, scoring='neg_mean_squared_error')
```

Output:
```
Negative Mean Squared Error for Each Fold: [-3637.06355263
-842.21947368 -3482.09421053]

Mean Negative Mean Squared Error: -2653.7924122807012

Standard Deviation of Negative Mean Squared Error:
1282.5368761792593
```

In machine learning, an imbalanced dataset refers to a dataset where one class (or more generally, a set of outcomes) is significantly underrepresented compared to others, especially in classification tasks. This disproportion in class distribution can introduce various challenges and biases in the training of machine learning models. Imbalance can lead to a biased model that prefers to make predictions favoring the majority class, as it's "easier" to achieve a higher accuracy by simply making predictions highly based on the majority class most of the time. This can be a serious issue, especially in scenarios where the minority class is the more important one to classify correctly. An illustrative example to provide is a machine learning model that can detect cancer from medical images: while cancerous patients would (hopefully) be less common in the dataset than healthy patient images, it is critically important that the model can identify cancer in future patient scans. As stated, in this example the minority class in the dataset would likely be images with cancer present, which could be as low as 5% of the data. So simply put: how do you steer the model away from achieving 95% accuracy in training by just predicting non-cancerous in every scenario, since 95% of the time that is the case.

In this example, the term "class" refers to the "Flow_Type" column, which has two values: "Laminar" and "Turbulent." We have significantly more samples that are laminar, making it the majority class. We can down-sample the majority class to balance the class distribution in a dataset by reducing the number of examples in the majority class. In the code below we can apply a random undersampling.

```
# Create a random under-sampler
rus = RandomUnderSampler(random_state=42)

# Apply the random under-sampler
X_downsampled, y_downsampled = rus.fit_resample(X_features,
y)
```

Output:
```
Class distribution before resampling: Counter({'Laminar':
47, 'Turbulent': 10})

Class distribution after resampling: Counter({'Laminar':
10, 'Turbulent': 10})
```

Another approach we can add to our playbook for class imbalance circumstances is the Synthetic Minority Over-sampling Technique (SMOTE, with the regression 'version' being called SMOTER). Unlike simple over/under sampling techniques, SMOTE creates synthetic samples (helpful to reduce overfitting). As in the undersampling example above, having only ten samples of both is clearly insufficient and instead creating more samples would be necessary. Generally, it starts by collecting random samples from the minority class and proceeds to find nearest neighbors (five is a typical number of neighbors). Then a new datapoint is created when interpolating between the original point and one of the neighboring points. This new synthetic point is placed on the line joining the original instance and the neighbor, and the process is repeated for new points to create a scatter of synthetic points (of minority class) in the feature space. While it may seem odd, I feel obligated to state that several high-ranking ML practitioners (e.g. Kaggle guild masters, active voices on LinkedIn and Twitter from MAANG companies, etc.) are quite enthusiastic in broadcasting their pessimism that SMOTE works 'at all'. The choice is yours.

Next, we can talk about regularization techniques, like L2 regularization. L2 regularization is particularly valuable when working with small datasets, as it can help prevent overfitting by adding a penalty term to the model's loss function (to keep the weight values small). Overfitting is a common problem with small data, where the model learns the noise in the training data, rather than the underlying pattern, leading to poor generalization on unseen data. L2 regularization works by adding the squared magnitude of the weights as a penalty to the loss function. By controlling the complexity of the model through the regularization parameter, it ensures that the model does not become excessively complex and fits the small data too well.

There are other techniques for regularization, as using L2 is just but one example. We can use the absolute values of the weights (L1 regularization), as a similar method of just modifying the terms in the loss function. However, other approaches like modifying the sampling or modifying the training algorithm also exist.

Now to address a very important concept for the last section of this chapter on data preparation: how to split the data into different groups as you begin to train and validate models. This will bleed perfectly into the next chapter, but feels more appropriate in this section, as it is an operation for preparing your data.

You will often hear the phrases "training data", "testing data", and "validation data". You will, unfortunately, probably also hear them misused. One thing that might not help is that different (famous) libraries define test and validation different, which is not incorrect but does cause confusion. Hopefully, we can set the record straight here on this simple but important concept.

After you have gone through the previous steps in the process to prepare your data, you will need to separate it into these three groups before beginning to train your model (usually by a random selection and dividing by percentages similar to 70/20/10). Training the model means that it will learn from the input data and adjust its internal parameters during training to minimize the error. An optimization process typically involves techniques like gradient descent, or similar, to effectively fit the data. This is not unlike the process of convergence for an engineering simulation (with loss being similar to residual).

The processes for setting these 'internal parameters' (weights, biases) is the process of training over iterations (epochs). At every epoch, the training data is used to propose (forward pass) what the coefficients should be and at the end of such training the validation data is used to evaluate the error. The process of training is usually comprised of many epochs (usually on the orders of 20-5,000) to reach acceptably low levels of error. Hence, the model has been exposed to both training and validation data to this point when determining the model setup/parameters. When it's all finished, the 'testing data' is used for 'blind' testing of the model performance. It's important that this test happens on a separate dataset (the testing set), which it has not seen during training to ensure an unbiased assessment. Note that sweeps for hyperparameter tuning should also happen prior to evaluation on the testing dataset. You can see a flowchart of this process in Figure 20. Sometimes, you'll hear 'evaluation dataset', which often also refers to the samples used for blind testing with what we just called the 'testing' dataset.

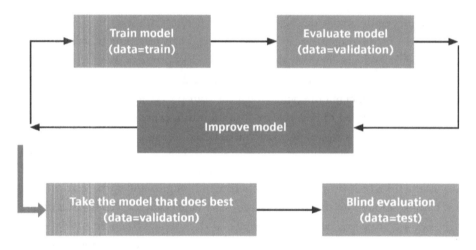

Figure 20: Illustration of how the training, validation and testing data is used in the process of creating a ML model.

Here is a simple snippet of code to illustrate this partitioning of datasets with the popular scikit-learn machine learning library. There are plenty of other considerations to build from this simple example: whether or not to shuffle the data before splitting, if you should use k-fold validation, or whether to use a random state or pass an integer to make reproducible, to name a few.

```
import numpy as np
from sklearn.model_selection import train_test_split
X, y = np.arange(10).reshape((5, 2)), range(5)

X_train, X_test, y_train, y_test = train_test_split(X, y,
test_size=0.33, random_state=42)
```

6.   PICKING AND TRAINING MODELS

The good news is that the training stage may be a more straightforward chapter in this text than the previous steps; problem/project definition, pre-processing, exploration, etc. As far as the machine learning pipeline this text is walking through, we're now ready to select different machine learning models and conduct trainings.

As mentioned before, it's not a bad idea to start with a linear regression model as a first model to train on; it can provide a quick (computationally efficient) prediction even on a large dataset with a lot of features, it's interpretable and arguably the most straightforward model to use, it can serve as a baseline for other models, and among other things, it can be less prone to overfitting. Additionally, I can make a few general recommendations which can help as a starting point for picking the right model for your problem. Disclaimer: this is definitely not intended to be a rule or absolute truth, every dataset is unique and can defy expectations regarding which models are best.

When working on engineering simulation data, I really like the cheat sheet by Monolith [10] which provides a flow chart with thoughts and steps to consider for learning about your dataset. After the 'start' -ing point, we can consider the right side for scenarios where we have tabular data, which could be a table(s) summarizing a number of different samples (simulations) for example. If we don't want to make new predictions, we can employ techniques for gaining further insights into our dataset. Compression techniques, like principal component analysis (PCA) which we covered as well as linear discriminate analysis (LDA), can extract some major underlying patterns in the data. LDA can help identify which inputs are related (helping us trim some out), and PCA can help us see strong trends even inside high-dimensional data. Parallel coordinate plots can also help show which variables are influential on the outputs, especially for highly multivariate data. Plotting the data in this manner is very cheap and can rapidly help in understanding relationships and spotting patterns or anomalies in the dataset. From these exercises, a few specific relationships between different variables may be emboldened to the practitioner as important.

While machine learning models are rigorous at capturing non-linear relationships, we can first try to fit a medley of simple and well-known functions to these specific variables we just identified (linear, polynomial of various orders, exponential, logarithmic, or power law, for example). We could also call it 'automatic correlation discovery', as simple algorithms are used to automatically sift through vast amounts of data to discover meaningful correlations without explicit human direction. We don't always need advanced machine learning models to provide correlations, and a quick screen with such an approach can perhaps illustrate simpler models do the job well!

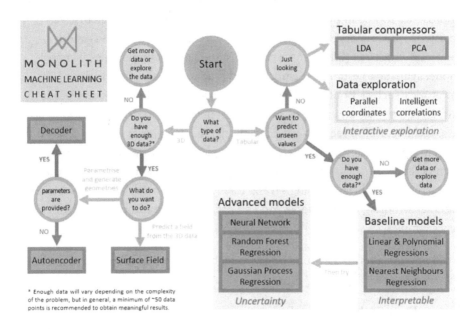

Figure 21: A roadmap for experiments and exercises when approaching a machine learning project. Permissions for sharing generously granted by Monolith AI [10].

To continue through the chart, we have the impervious question of 'do we have enough data' up next. Naturally, the implicit context of the question really makes the question more completely stated as 'do we have enough data for the problem we wish to understand', since we don't care about all relationships of the data. It's great to ask the question in this way because it implies that, whether we have enough data or not, there is a screening step to confirm our data is suitable to confidently build an accurate and general model in the bounds of the problem we wish to solve. You'll often hear about the trade-off at hand in making a model that is the right balance of overfit (so it can be accurate enough on the training data) and general (so it can perform well on new and unseen data). One cannot make claims to understand the nature of their data in this regard without performing some form of analysis and exploration to understand it in the bounds/range of their interest. Hence why I view 'do I have enough data' as a very related question to 'how general is my machine learning model'.

Since we have talked a lot about interpretability, I will go straight into discussing uncertainty from the 'advanced models' block in the diagram. Typically, the regression models we use provide point estimates, but they don't capture the uncertainty around these predictions. In the realm of uncertainty quantification in machine learning, we have options on how to estimate uncertainty: there are approaches rooted in the intrinsic properties of specific models themselves (epistemic uncertainty), such as Bayesian Neural Networks (BNNs) and Gaussian Processes (GPs), or we can use systematic methodologies that are model agnostic to estimate the uncertainty/variance in the observations (aleatoric), such as Bootstrap Ensembles and Monte Carlo Dropout. For the latter, generally we are trying to vary 'things' pertaining to the model and data (e.g. architecture) to see how the predictions/accuracy varies (we can refer to this loosely as variance). Specifically, a few examples of things we could vary and then evaluate the accuracy in the predictions are dataset sub-samples we draw (bootstrap ensembles), training several totally separate models or architectures (deep ensembles), turning nodes 'on and off' at random (dropout), to name a few examples.

I will provide a few paragraphs below to highlight some quick information you may find useful.

TensorFlow Probability. A Python library built on TensorFlow that makes it easy to combine probabilistic models and deep learning. For example, it offers layers like 'DenseVariational' for BNNs.

PyTorch & Captum [11]. Captum is a model interpretability library for PyTorch. It provides uncertainty estimates using techniques like Integrated Gradients. It can work on any PyTorch model.

Scikit-learn. For classifiers that support the 'predict_proba 'method (e.g., logistic regression, random forest classifier, gradient boosting classifier), you can use this method to get the probability estimates for each class. For regression, ensemble methods can provide variance estimates. If a classifier does not natively support the estimation of class probabilities, you can use the 'CalibratedClassifierCV' to calibrate the classifier and obtain probability estimates.

Concerning the left side of the diagram tree in Figure 21 regarding 3D data; I will only provide some comments and then refer to other sections of the book because the verbiage/terms are tailored to Monolith's product specifically. Firstly, the '3D' data-type formats can potentially refer to the geometry, mesh, and sometimes even the boundary conditions or properties of the simulations. This is the holistic set of information we need to collectively describe the actual field data results (which would be 3D in nature), as well as the associated meta data parameters (which could describe the operating conditions associated to the 3D field data). Think about a simple example of 3D data fields for a batch of different simulations of flow over a cylinder: we need the 3D fields but we also need input parameters (meta data) which may describe the variations between the simulations (e.g. Reynolds number, freestream turbulence intensity, and cylinder diameter). Referring to an earlier section in this text, file formats such as cgns, vtk, and stl would be commonplace. Coming (not much) later in this text will be a review of different models, whereby the ones relevant for 3D data will be overviewed (variational autoencoders, convolutional neural networks, and graph neural networks). Please refer to the writeup for these sections to learn more about the models themselves.

In addition, I also really like to refer to the flowchart image by scikit-learn [12] to inspire a framework for approaching various problems. While scikit-learn is widely used and is absolutely not specific to engineering data or the CAE-space, their flowchart helps the user to find reasonable models for their specific datasets to pick as a starting point. It's again stated that these numbers cannot be considered absolute or as a 'one size fits all' recommendation. I am going to put the figure on the next page so that it looks properly readable, given that it's a large figure and could be hard to see if shrunk.

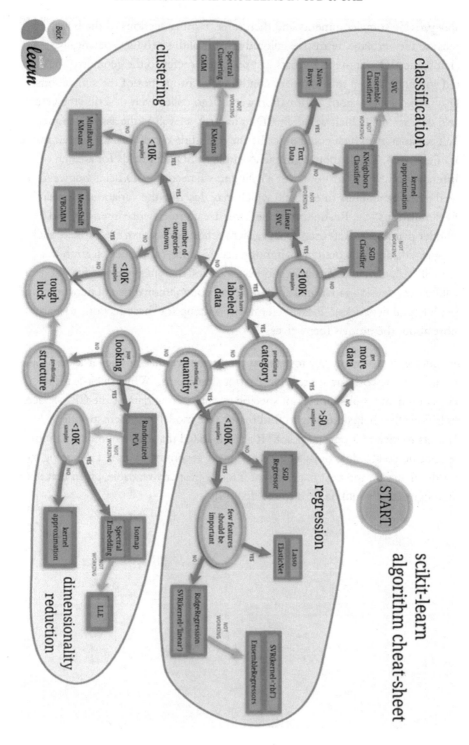

Now that some general remarks have been made regarding a framework you can adopt to try different models, we can proceed to highlight some different machine learning models with a small description of each. We will only cover models of relevance to the CAE industry (and even so, there will be more we won't cover).

"STANDARD" NEURAL NETWORKS

Neural networks are a cornerstone of machine learning, inspired by the structure and functions of the human brain. They consist of layers of interconnected nodes, or 'neurons,' each performing simple calculations. Through a process called learning, these networks adjust their internal parameters in order to create a model that can make accurate predictions based on input data. Since an overview of the mathematics and the calculations taking place under the hood are covered in innumerable textbooks and web pages, I don't feel this book should dwell on the topic. To (try to) make a (somewhat) useful contribution, I will only hit a few concepts and then move on. Warning: I will be quirky to involve unique resources I think aren't well known (but they are fun).

Let's start by analyzing a single neuron in the diagram shown in Figure 22 to illustrate the calculation that takes place for just this little piece of the network. This neuron will receive multiple inputs, as it will be fully connected to all neurons in the previous layer. Each input is associated with a weight, which is a parameter that signifies the importance or strength of the input to the neuron's output. If a neuron has n inputs, denoted as $x_1, x_2, ..., x_n$, and corresponding weights $w_1, w_2, ..., w_n$, then the weighted sum of the inputs times their respective weights is calculated.

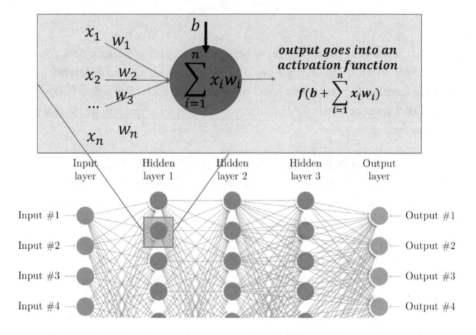

Figure 22: Sample of mathematical operations associated with a neuron in an arbitrary network.

A neuron also has a bias term, b, which is added to the weighted sum. The bias allows the neuron to shift the activation function to the left or right as another degree of freedom, which can be critical for successful learning.

Finally, the total input (the things we have talked about so far) is passed through an activation function. The activation function is a non-linear transformation that decides whether a neuron should be activated or not. Common activation functions include the sigmoid function, the hyperbolic tangent (tanh), and the Rectified Linear Unit (ReLU). The final output of the neuron is what value comes out of the activation function. Fun fact; classically, the sigmoid and hyperbolic tangent activation functions are biologically motivated, since our neurons do respond similarly based on the signal they receive. However, plenty of others are used that are not inspired by our biology, such as physics-based activation functions. Each of these neurons are stacked together to build a sequence of calculations that builds an output from the inputs.

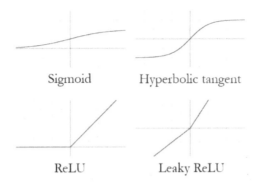

Figure 23: A sample of 4 activation functions. X-axis is the input to the activation function (weighted sum of the inputs plus a bias term), and y-axis is the output of the activation function.

The entire process can be viewed as a two-step transformation where the linear combination of inputs (weighted sum plus bias) is first calculated, and then a non-linear function is applied to this sum. This structure allows the neuron to learn complex patterns in the data, and when many such neurons are connected together in a network, they can model even more complex relationships.

So informally stated, at its heart it is just a nested composition of functions with some weighting to say how strongly different neurons should be connected. Again, these weights that are assigned are a result of the training process where you use an optimization algorithm to assign their ideal values in order to minimize some error. You may initialize with some random weights or values for all the neurons, but during training you will continually update the weights until your model has converged (finished training).

The training of such networks, especially large ones, is computationally intensive, often involving the simultaneous updating of millions or billions of parameters. This is where GPUs and parallel computing come into play, as well as acceleration libraires (like CUDA, which is why a fresh computer setup always begins with a CUDA installation when setting up your machine learning environment). GPUs are particularly well-suited for neural network training due to their ability to perform a large number of floating-point operations in parallel, significantly speeding up the calculations involved in both forward and backward passes of the network. Each neuron's operations can often be computed in parallel, and likewise, many aspects of the training process, like batch processing and gradient calculations, are parallelizable. These are small calculations individually, which are handled by threads on the GPU. You can search for any

GPU's 'CUDA cores' count on the internet to get an idea on how many parallel channels there are for your calculation.

With that brief summary concluded, let's get you started with creating your first neural network! There are many different ways to do this, so we will cover two equivalent approaches using popular libraries: Keras and PyTorch. Both Keras and PyTorch are used to help streamline building and training machine learning models. They provide a high-level abstraction for this process, which means that it will be easy for users to create sophisticated models without getting their hands dirty in the underlying mathematics or exhaustive coding.

In 2017, Google announced that Keras would be integrated into TensorFlow as its official high-level API, starting from TensorFlow version 1.2. This integration meant that while you could still use Keras with other backends, its development would be more tightly integrated with TensorFlow (so think of the two as in the same category). While on the other hand PyTorch is developed by Facebook's AI Research lab, as an open-source machine learning library also written in Python.

The dataset we will use will be a famous dataset for identifying onset of diabetes in Pima Indian populations (this can be found in several places, such as UCI Machine Learning repository). The dataset can be used for predicting whether an individual will experience diabetes onset, or not. The input variables are numerical and provide information about each patient towards this classification of diabetes.

Let's start with the Keras way of making our neural network. First, let's import needed libraries for the project.

```
#Import libraries
import numpy as np
from numpy import loadtxt
from tensorflow.keras.models import Sequential
from tensorflow.keras.layers import Dense
from sklearn.metrics import confusion_matrix,
classification_report, roc_curve, auc
import pandas as pd
import urllib.request
import matplotlib.pyplot as plt
```

For the reader's convenience, let's start by importing the dataset from a URL rather than expecting the user to download the file and local import it as a .csv file. After using the 'urllib.request' library to grab the data, we will also split the inputs and outputs of the machine learning model.

```
# URL of the dataset
url =
"https://raw.githubusercontent.com/jbrownlee/Datasets/maste
r/pima-indians-diabetes.data.csv"

# Open the URL and load the dataset
with urllib.request.urlopen(url) as response:
    dataset = np.loadtxt(response, delimiter=',')

# Displaying the first few rows of the dataset to verify
print(dataset[:5])

# split into input (X) and output (y) variables
X = dataset[:,0:8]
y = dataset[:,8]
```

In Keras, we will define our model layer by layer, in a consecutive sequence of code lines. We will create a Sequential model to add our layers, and we have flexibility here to create an architecture that we are happy with.

```
# define the keras model
model = Sequential()
model.add(Dense(12, input_shape=(8,), activation='relu'))
#expects 8 variables as input, as 12 neurons in following
layer
model.add(Dense(8, activation='relu')) #subsequent layer
has 8 neurons
model.add(Dense(1, activation='sigmoid')) #output has 1
neuron, since it's just predicting onset of diabetes 'yes'
or 'no'
```

Since we have defined our model, it is now time to compile it. Compiling the model uses the efficient numerical libraries under the covers (the so-called backend) such as Theano or TensorFlow [16]. This process will require us to specify a few extra pieces of information, which can be as extensive (or brief) as the number of things you want to specify. Some minimum things need to be specified though, like the loss function (in this case using binary cross entropy), the optimizer that is used during training to set the coefficients to ideal values (weights and biases), and the metric we want to report during training to tell us how accurate the model fitting is going.

```
# compile the keras model
model.compile(loss='binary_crossentropy', optimizer='adam',
metrics=['accuracy'])
```

The next code segment will be training the Keras neural network model using the input data X and the target labels y, as we iterate through the dataset samples 150 times (epochs). It processes the data in batches of 10 samples each time, updating the model's weights to minimize the loss and improve accuracy with each epoch. You can train your model on your loaded data by calling the fit() function on the model.

```
# fit the keras model on the dataset
model.fit(X, y, epochs=150, batch_size=10)

Output:
Epoch 1/150 77/77 [==============================] - 1s
2ms/step - loss: 7.4059 - accuracy: 0.6458

Epoch 2/150 77/77 [==============================] - 0s
2ms/step - loss: 2.2404 - accuracy: 0.6094

Epoch 3/150 77/77 [==============================] - 0s
2ms/step - loss: 0.9417 - accuracy: 0.5833

...
```

Now that our model is trained, we want to start extracting insights from it regarding its accuracy and predictive capabilities. We can start with a simple prediction of accuracy, which evaluates the performance of the trained Keras neural network model on the same dataset it was trained on. We can define accuracy as the proportion of correctly predicted instances out of the total instances in the dataset (whereby '24' indicates the evaluation was conducted on 24 batches of data samples). Frankly, you can define accuracy as whatever you want, which could be your silver bullet in your machine learning project if you intuitively identify the right metric to use. Whether I am working on FEA simulation data for Von Mises stresses, or flow-field information in a CFD simulation, the choice should be tailored to each project you are working on in order to maximally pick the right term which can be optimized by your machine learning model so that the actual prediction quality improves.

```
# evaluate the keras model
_, accuracy = model.evaluate(X, y)
print('Accuracy: %.2f' % (accuracy*100))
```

Output:
```
24/24 [==============================] - 0s 2ms/step -
loss: 0.4874 - accuracy: 0.7695 Accuracy: 76.95
```

To get a bit more information back to us regarding the accuracy of the predictions, we can form a side-by-side comparison of 'predicted versus actual', while also changing the binary nature of the predictions to that of a number as a form of probability (which can give us a means of confidence, since for example 0.51 versus 0.99 may both round to '1' to indicate diabetes onset, but are significantly different when we cast it as 0.51 versus 0.99).

```
# make probability predictions with the model
predictions = model.predict(X)
# round predictions
rounded = [round(x[0]) for x in predictions]
combined_matrix = np.column_stack((predictions, y))
print(combined_matrix[0:10])
```

Output:
```
24/24 [==============================] - 0s 1ms/step
[[0.5826624  1.        ]
 [0.09126351 0.        ]
 [0.84923083 1.        ]
 [0.10615474 0.        ]
```

```
[0.85944366 1.          ]
[0.25273097 0.          ]
[0.18787168 1.          ]
[0.63791072 0.          ]
[0.99205285 1.          ]
[0.08115277 1.          ]]
```

Next, we will create a confusion matrix for our diabetes onset classification. These are very common for classification projects, as the matrix helps succinctly summarize to us the tendency for our model to make different types of prediction errors. The matrix is just a summary of the true positive (TP), true negative (TN), false positive (FP), and false negative (FN) predictions by count for each. You can and should read about this more thoroughly if classification problems are a priority to you.

```
# Confusion Matrix
cm = confusion_matrix(y, rounded)
print("Confusion Matrix:")
print(cm)
```

Lastly, before making a plot, we can take advantage of more pre-canned report to print with different accuracy metrics with 'classification_report'. This text will provide more information on these specific metrics elsewhere, but in short they are the accuracy of positive predictions, a sensitivity, and balance of precision and recall (from left to right).

```
# Classification Report
cr = classification_report(y, rounded_predictions)
print("Classification Report:")
print(cr)
```

Output:
```
Classification Report:
              precision    recall  f1-score   support

         0.0       0.79      0.88      0.83       500
         1.0       0.72      0.56      0.63       268
```

			0.77	768
accuracy			0.77	768
macro avg	0.75	0.72	0.73	768
weighted avg	0.76	0.77	0.76	768

Finally, we will print a plot and a score; the AUC (Area Under the Curve) score is a measure of the model's ability to distinguish between the positive and negative classes. An AUC score of 1 represents a perfect classifier, while a score of 0.5 represents a worthless classifier.

```
# ROC Curve and AUC
fpr, tpr, thresholds = roc_curve(y, predictions)
roc_auc = auc(fpr, tpr)

plt.figure()
plt.plot(fpr, tpr, color='darkorange', lw=2, label='ROC
curve (area = %0.2f)' % roc_auc)
plt.plot([0, 1], [0, 1], color='navy', lw=2, linestyle='--
')
plt.xlim([0.0, 1.0])
plt.ylim([0.0, 1.05])
plt.xlabel('False Positive Rate')
plt.ylabel('True Positive Rate')
plt.title('Receiver Operating Characteristic')
plt.legend(loc="lower right")
plt.show()
```

Output:

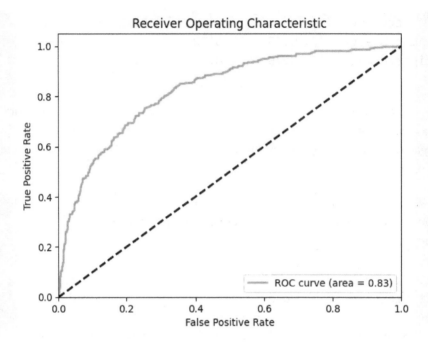

While not a result, we showcase a simple snippet that will print a summary of your neural network. This is a simple print of the model architecture (number of layers, type of each layer, output shape of each layer, and number of parameters total in each layer). This is especially helpful if you are constructing complicated networks that are large and have errors related to mismatch in shapes. For example, when you have a 'bottleneck' in your neural network, like a Variational Autoencoder, that varies the neuron count per layer from high-to-low to high again.

```
# Model Summary
print(model.summary())
```

Output:
```
Model: "sequential_2"
```

```
 Layer (type)                Output Shape
 Param #
=================================================================
======
 dense_6 (Dense)             (None, 12)                    108
```

```
dense_7 (Dense)              (None, 8)                 104

dense_8 (Dense)              (None, 1)                 9

=========================================================
Total params: 221 (884.00 Byte)
Trainable params: 221 (884.00 Byte)
Non-trainable params: 0 (0.00 Byte)
```

As promised, let's move to repeating the same problem and neural network model implementation but with PyTorch instead of Keras. We don't need to provide the full code, as some steps are almost identical to the previous demonstration. But we are going to certainly highlight the different steps needed for PyTorch, specifically.

Firstly, after importing the data, the data (which was NumPy arrays) needs to be converted to PyTorch tensors. You'll find this is a common step in projects with PyTorch, and that conversion early on in projects can avoid mismatched data formats later, which usually would result in an error and interrupt your code execution. You can perform this conversion by creating new tensors out of existing NumPy arrays.

```
X = torch.tensor(X, dtype=torch.float32)
y = torch.tensor(y, dtype=torch.float32).reshape(-1, 1)
```

Similar to before, we create a Sequential model with the layers listed. The first layer indeed needs to match the input data, and you can notice the sizes for each part of the network matches the previous code in Keras by consisting of three linear layers interspersed with ReLU activation functions and ending with a sigmoid activation function. This is suitable for a classification problem due to the sigmoid activation function, which outputs a value between 0 and 1, which is ideal as the value can be interpreted as a probability.

```
model = nn.Sequential(
```

```
        nn.Linear(8, 12),
        nn.ReLU(),
        nn.Linear(12, 8),
        nn.ReLU(),
        nn.Linear(8, 1),
        nn.Sigmoid()
)
print(model)
```

Output:
```
Sequential(
    (0): Linear(in_features=8, out_features=12, bias=True)
    (1): ReLU()
    (2): Linear(in_features=12, out_features=8, bias=True)
    (3): ReLU()
    (4): Linear(in_features=8, out_features=1, bias=True)
    (5): Sigmoid()
)
```

Now, we can see more significant differences in the code below. Not so much the definition of loss, optimizer, batch size, and epochs, but indeed the way training is defined. PyTorch offers more explicit control over the training loop, requiring you to manually iterate over epochs and batches. This allows for more customization but also requires more code. Keras, being a high-level API, abstracts away many of these details, making the code more concise. The training loop is handled internally by the fit method, reducing the amount of boilerplate code needed.

```
loss_fn   = nn.BCELoss()  # binary cross entropy loss
optimizer = optim.Adam(model.parameters(), lr=0.001)


n_epochs = 100
batch_size = 10


for epoch in range(n_epochs):
    for i in range(0, len(X), batch_size):
        Xbatch = X[i:i+batch_size]
```

94

```
    y_pred = model(Xbatch)
    ybatch = y[i:i+batch_size]
    loss = loss_fn(y_pred, ybatch)
    optimizer.zero_grad()
    loss.backward()
    optimizer.step()
print(f'Finished epoch {epoch}, latest loss {loss}')
```

Output:
```
Finished epoch 0, latest loss 0.731421947479248
Finished epoch 1, latest loss 0.7107272744178772
Finished epoch 2, latest loss 0.6836671233177185
Finished epoch 3, latest loss 0.6536120772361755
```

Neural networks can tackle a wide variety of problems, and as such have a wide variety of architectures that you may see in lots of different applications. Certain variations have grown quite popular and emerged as a distinguished type of neural network (like graph neural network, or autoencoder). While the 'vanilla' or 'standard' neural network may stick in your brain as a resemblance of Figure 9, it can go by several names that usually have terms like 'fully connected' and 'feed-forward', or even be called 'artificial neural network (ANN)'. To avoid confusion, let's mention a few other types of neural networks now so you can see the bigger landscape. I have had many conversations with confused individuals not realizing that many machine learning models are still in the neural network family just because of their name! For example, did you know that Large Language Models (LLMs) like ChatGPT are indeed a type of neural network. Specifically, they belong to a family of neural networks called "transformers," which are designed for handling sequential data, such as text! Some of these will have dedicated sections to follow, but as an overview let's make a short list.

Standard Neural Networks (ANNs): These are the most basic form, typically made of fully connected layers. They are suited for a wide range of predictive tasks and look akin to aforementioned figures.

Convolutional Neural Networks (CNNs): Specialized for processing data with a grid-like topology, such as images. CNNs use convolutional layers to capture spatial structures/patterns in data, making them powerful for tasks like image classification and object detection.

Recurrent Neural Networks (RNNs): Designed for sequential data, like time series or text (though transformers are now often preferred for many natural language processing (NLP) tasks on text data). RNNs have the unique feature of maintaining a 'memory' of previous inputs in their hidden layers, enabling them a competence in handling temporal behavior. Essentially, the output from one step is the input to the next step. Long Short-Term Memory Networks (LSTMs) is a famous type of RNN that can learn long-term dependencies in data sequences.

Autoencoders: (Mostly) used for unsupervised learning, particularly in dimensionality reduction and feature learning. Autoencoders learn to compress data into a lower-dimensional representation and then reconstruct it back to the original form. Similar to PCA/POD, but with non-linear capabilities for capturing patterns. *Variational Autoencoders (VAE)* are a special kind of Autoencoder and particularly famous at the moment in the 'generative AI' space, as they can produce new data after learning a latent space.

Generative Adversarial Networks (GANs): Consist of two neural networks, the generator and the discriminator, that are trained simultaneously in a game-theoretic framework. GANs are known for generating high-quality synthetic data, especially images.

Graph Neural Networks (GNNs): Designed to process data represented as graphs. GNNs are effective in tasks that involve relational reasoning and graph-structured data, like social network analysis and molecular biology. For CAE, GNNs are useful in capturing trends and phenomena that occur over a broad spectrum of scales (e.g. length, time), not to mention they are suitable to work directly on the discretized meshes simulation practitioners create.

Bayesian Neural Networks (BNNs): Incorporate principles of Bayesian statistics to quantify uncertainty in the predictions. They are useful in scenarios where uncertainty estimation is valued, as well as scenarios where data scarcity is at play- which is frequently the case in simulation!

Transformer Networks: Originating from the field of natural language processing, transformers have a unique attention mechanism that allows them to process sequences of data in parallel, significantly improving efficiency and effectiveness in tasks like language translation and text generation. This is type of neural network is booming the last two years in the realm of generative AI. For example, ChatGPT is a pre-trained model that uses a transformer architecture (which of course generates text akin to a human conversation).

Now that we've provided this topical survey, let's cover a few types of neural networks.

CONVOLUTIONAL NEURAL NETWORKS (CNNS)

Convolutional neural networks (CNNs) are great for finding patterns in images to recognize objects, and thus have been widely successful in applications like image classification, object detection (e.g. drawing bounding boxes around object), image segmentation (e.g. distinguishing between tissue types in medical imaging), facial recognition, and more. The number of industries successfully taking advantage of CNNs feels innumerous, but to name a few: healthcare and medical imaging, transportation (autonomous vehicles), security and surveillance, manufacturing and quality control, agriculture, entertainment, and media, among others.

Let's review a diagram of a sample CNN by LeCun [13] to provide a basic understanding of how CNNs work. This will start from the input data (image) and then work through the subsequent layers in the model that process the data in steps.

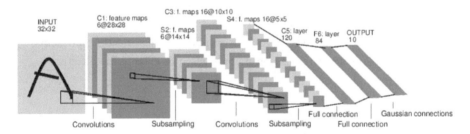

Input Image:

This is the initial image that you want to classify or analyze using the CNN. The image is represented as a matrix of pixel values, where each pixel can have multiple channels (e.g., red, green, blue for color images). In this case, it's the letter 'A' with 32x32 pixels (which is small).

Convolutional Layer:

This layer applies convolution operations using filters (or kernels, which are little squares that 'scan' the full image from left to right and top to bottom) on the input image (or to the output from the previous layer). This is the magic of CNNs. The goal of the convolution operation is to detect local patterns such as edges, textures, and other features. Multiple filters are applied, resulting in multiple feature maps as outputs. The size of the feature maps can be controlled by the stride (how many pixels the kernel slides) and padding (empty pixels around border of image) used in the convolution operation.

Activation Function:

After the convolution operation, an activation function (typically the Rectified Linear Unit (ReLU) or hyperbolic tangent), is applied to introduce non-linearity to the model.

Pooling Layer (or Subsampling):

This layer reduces the spatial dimensions of the feature maps, preserving the important features. Pooling helps reduce the computational load and provides a form of translation invariance. The most common method is max pooling, where only the maximum value in a local patch of the feature map is kept. While it may be counterintuitive, since it is lowering the resolution, it is often apparent that this helps improve the predictive capabilities of the neural network. As you move to the right of the diagram, with more and more subsampling layers, it eventually gets quite small until you are dealing with very condensed layers. These condensed layers (covered below, but spoiler here) are essentially taking all the features learned from the previous layers and (for classification problems) doing almost a logistic regression on them to learn to classify the different outputs (different letters).

Fully Connected (FC) Layers:

After several convolutional and pooling layers, the high-level features extracted are flattened into a one-dimensional vector and fed into one or more fully connected layers. These layers are traditional neural network layers where every node is connected to every node in the subsequent layer. The purpose of these layers is to combine the features extracted by the convolutional layers to make a final prediction.

Output Layer:

The final fully connected layer produces the desired number of outputs. For classification tasks, the number of nodes in the output layer corresponds to the number of classes. A softmax activation function is often applied to produce a probability distribution over the classes.

I would like to shift the discussion now to reference a code project as a starter tutorial [14] with the following objective: predicting a coefficient value from a CNN model that has ingested image data (CT scans, which are stacked 2D slices to represent a 3D volume). The coefficient value is used in clinical settings to describe the volume of air exhaled by a person in one full breath. Ordinarily this is used to baseline healthy lungs.

I chose this code/example after reviewing dozens of examples online: it is a regression problem, it has well documented and discussed code you can run for free on Kaggle's cloud, you can see different models applied, you can see different pre-processing and augmentation methods applied in various experiments (e.g. EDA), and most of all you can parallel this example for CAE applications. For example, this problem formulation is similar to an external aerodynamics project in automotive: the data is three-dimensional, there are many samples (simulations being a parallel to patient's imaged) the model needs to manage, both small and large scales of characteristics in the individual scans matter towards the final result (e.g. small flow separations as well as large patterns in the flow-field), conversion of local information in images to a single regression value (drag coefficient), among others. There are many models, notebooks, and steps implemented in this project, but I will just provide some code concerning a specific model and manipulations you can use, as well as some high-level steps for the project.

First, to no surprise, we import a few libraries critical and provide some comments on their general purpose for us:

```
import torch; import torchvision #For model and dataloader creation
import pytorch_lightning as pl #For efficient and easy training implementation

#ModelCheckpoint and TensorboardLogger for checkpoint saving and logging
from pytorch_lightning.callbacks import ModelCheckpoint
from pytorch_lightning.loggers import TensorBoardLogger

import imgaug.augmenters as iaa #imgaug for augmentation
```

```
pipeline
```

```
#(Many others, just a subset of libraries shown that are
needed)
```

Next, we cover code that defines a set of image augmentation operations that will be applied to the training images. We are going to totally skip a huge pre-processing step which would be imperative to the success of the resulting model: building the dataset, resizing/modifying as needed, exploring the images to spot any areas for improvement, any sort of normalization, dividing or training-validation-testing, and others. This is a dense and detailed area for learning, so I recommend exploring the reference provided for this project to see various discussions and experiments when evaluating these individual steps.

Data augmentation is a set of techniques used to artificially increase the size of the training dataset by applying various image transformations. This can help improve the model's performance and its ability to generalize to new data, as the transforms help make the model more robust to variations in the incoming data. There are so many more possibly ways to do this, but again, I want to get the reader acquainted with a basic approach so they can feel more comfortable jumping into projects.

```
# We create the dataset objects and the augmentation
parameters
train_transforms = iaa.Sequential([
                        iaa.GammaContrast(),
                        iaa.Affine(
                            scale=(0.8, 1.2),
                            rotate=(-10, 10),
                            translate_px=(-10, 10)
                        )
                    ])
```

Next, we specify a few important things related to the 'data loaders' in PyTorch, which can make stepping through loading datasets during training/validation efficient and clean. Firstly, we specify parameters to define how many samples are loaded and processed concurrently during model training or evaluation (batch size). Additionally, we specify the number of workers, which are CPU subprocesses, to use for loading the data (reading the data, applying transforms, and sending to training). Both of these choices are important to balance making the process go quickly (via parallelism) while staying within your computational resource available. Lastly, we define the loaders which use the training/validation dataset object, with these aforementioned parameters included which we just defined.

```python
# Adapt batch size & num_workers according to your
hardware.
batch_size = 8
num_workers = 4

train_loader = torch.utils.data.DataLoader(train_dataset,
batch_size=batch_size,

 num_workers=num_workers, shuffle=True)
val_loader = torch.utils.data.DataLoader(val_dataset,
batch_size=batch_size, num_workers=num_workers,
shuffle=False)
```

Next up, we would start to define a model architecture and set it up for training. There are a few steps/decisions to start with, but first let's talk about the convolutional base. Here's an example:

```python
model = models.Sequential()
model.add(layers.Conv2D(32, (3, 3), activation='relu',
input_shape=(32, 32, 3)))
model.add(layers.MaxPooling2D((2, 2)))
model.add(layers.Conv2D(64, (3, 3), activation='relu'))
model.add(layers.MaxPooling2D((2, 2)))
model.add(layers.Conv2D(64, (3, 3), activation='relu'))
```

To break this down: Conv2D is a two-dimensional convolutional layer. 32 is the number of filters (or kernels), which slides over the image in the shape of a (3,3). For the first layer, you need to specify the shape of the input data, which in this case is (32, 32) pixels and also has three channels (RGB). We can see the subsequent lines with Conv2D are additional convolutional layers with 64 filters each, also of size (3,3). In between these lines you'll see a max-pooling operation, which as stated above is a down-sampling operation, and in this case is using a (2, 2) window.

One thing that is particularly useful in projects, when trying to quickly experiment with several models, is calling pre-trained models from the public domain that are well known to be best in class for that problem (e.g. breast cancer detection). That way you do not have to start from scratch every time and can even use a few lines of code to tweak the pre-trained models for your application. I would like to provide a short number of lines of code to showcase a few possibilities in this area, and then move on. I will do this in TensorFlow to maintain a variety from the already shown PyTorch code.

First, we can instantiate the 'B5' variant of the EfficientNet architecture, which returns a Keras image classification model that can be loaded with weights pre-trained on ImageNet [15]. The inclusion of the B5 model is arbitrary for illustration purposes, and you could freely use other versions with customization.

```
tf.keras.applications.efficientnet.EfficientNetB5(
    include_top=True,
    weights='imagenet',
    input_tensor=None,
    input_shape=None,
    pooling=None,
    classes=1000,
    classifier_activation='softmax',
    **kwargs
)
```

Alternatively, we could apply transfer learning with the EfficientNetB5 model for a regression prediction.

```
import tensorflow as tf
from tensorflow.keras.applications import EfficientNetB0

# 1. Load the EfficientNetB0 model with pre-trained weights
base_model = EfficientNetB0(include_top=False,
weights='imagenet', input_shape=(224, 224, 3))

# 2. Freeze all layers of the base model for transfer
```

```
learning
for layer in base_model.layers:
    layer.trainable = False

# 3. Create a custom head for regression
x = base_model.output
x = tf.keras.layers.GlobalAveragePooling2D()(x)
x = tf.keras.layers.Dense(128, activation='relu')(x)
output = tf.keras.layers.Dense(1)(x)   # Single output
neuron for regression

# 4. Combine the base model and the custom head
model = tf.keras.models.Model(inputs=base_model.input,
outputs=output)

# 5. Compile the model for regression
model.compile(optimizer='adam', loss='mean_squared_error',
metrics=['mae'])
```

Thanks to their success in so many industries, there are a wealth of successful CNN architectures to pick from over years of development and innovation in the field. Here are a few names to leave you with as 'some of my favorites': ResNet50 (number can vary, larger number means more layers), EfficientNet (also called EffNets), DenseNet, and U-Net (notable variation: V-Net).

The above lines of code are just a set of starter examples to help you feel more equipped to jump into your next project. It would be far too extensive for the purpose of this book to include a realistic and thorough set of steps to cover many of the steps you may need to take in your CNN project to boost accuracy. To help get your feet wet, I would like to provide a succinct summary of tools and operations you could consider when getting deep into your future CNN projects.

- Setting a standard orientation (and/or physical dimensions) for all the images in your dataset
- Setting a standard physical resolution size so that each pixel of a data array represents the same physical size. You may need to do resampling, pad, and/or crop to achieve this
- Denoising your image data

103

- Setting a standard for the intensity values in your images
- Data format conversion for easy access (e.g. Nifti) and efficient data storage in your pipelines (e.g. npy, hdf5)
- Complex artifact corrections (e.g. maybe someone took a blurry picture, or captured other things in the image frame they didn't mean to)
- Quantification (changing units)
- Balancing the classes in your data
- There are many other augmentations you can make to the data/images to make the model more accurate. You can make new samples based on the original dataset via rotating, flipping, applying zoom, shifting, or shear-transformations to the images.
- You can also consider architecture enhancements, such as increasing the layer (depth) count of the network, incorporating skip connections to bypass layers (residual connections), batch normalization, or dilated convolutions, as a few examples

As stated, CNNs have a resounding history of success for images. There is some similarity/opportunity in CFD to take advantage of these advances in CNN algorithms and modeling capabilities, since sometimes our CFD data can be structured and grid-like which resembles image data that is used as input. However, these models are data hungry and often impose a soft limitation around 512x512 resolution images before applying CNNs becomes difficult. The amount of memory needed grows quadratically with the resolution, which can be a limiting factor, especially on GPUs. To manage this, many practitioners will down-sample high-resolution images to a size manageable for their CNN and hardware. However, CFD (for example) can have many more datapoints in an image (or array) for a single simulation, which may be 100M cells distributed in three dimensions if it is an industrial realistic simulation. Therefore, there have been many adaptions from mainstream CNNs into CAE-specific domain. You'll see some cases below in the VAE section for example, whereby 3D convolutional filtering can be down to abstract meaning artifacts in the flow-field through it's down-sampling actions. As of now, in early 2024, I anticipate GNNs and other methods might overtake some CAE-based applications for which CNNs were previously used.

AUTOENCODERS

One particularly special type of neural network is a variational autoencoder (VAE). Dimensionality reduction, anomaly detection, and generative modeling are some mainstream use cases you may hear about using VAEs for. In the CAE space, I believe they bring a lot of value so I will do my best to convey that without belaboring here about them for too long. VAEs can work with experimental data, both high or low fidelity simulation data, or even results originating from complicated non-linear systems that we seek simplified models of (e.g. ROMs) for design. Some benefits are enabling rapid (on the order of 5 minutes) optimization cycles for design by replacing expensive simulation with these reduced order models, gaining insight into underlying behavior of a set of variables, dimensionality reduction, and for cases in applying VAEs to surface meshes (stl files) they can even be used to create a parameterization for meshes/CAD which previously had no parameterization. From these capabilities listed, autoencoders are one concrete technology enabling generative AI/generative engineering (not just hype, but reported by Gartner research).

VAEs are comprised of several 'blocks' or components. The first is an encoder, which compresses the input data into a lower dimension, which I will call a latent space. For example, if you have 20 design variables to describe your possible design-space, then a VAE could reduce that down to just 7 variables (whereby all designs possible from the 20-design variable configuration could be replicated in that of the 7). Typically, your input data (whether stl files, or tabular data) may be geometric variations (for example) in your simulation work, which could be comprised of a design space of many variables to describe the geometry (for example 20 parameters would not be uncommon). An immediate advantage of a latent space you create with VAEs is it could hopefully be far fewer variables, for example 7 rather than 20, which means that sampling your design space with simulations is much cheaper! For example, if you want to do 3 variations of each parameter in both cases, then the 20 variable scenario requires 8,000 simulations whereas the 7 variable latent space would only require 343. The other main component in a VAE is the decoder, which is responsible for using the latent space to reconstruct the original data with as little error as possible between the two. Here is a simple image I made below in Figure 24 to show the setup to learn a geometry-space with a VAE for marine hulls, created for the purpose of doing a hydrodynamic study on a marine hull design project.

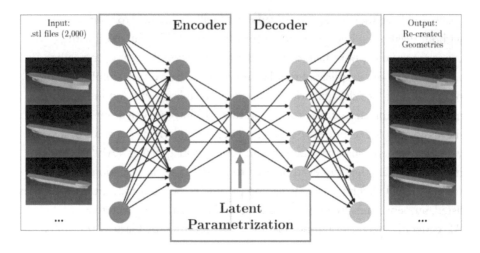

Figure 24: Flow of data, in this case .stl files describing the hull of a marine vessel with various geometries, from the encoder to latent space (compressed parameterization) to the decoder.

Here are a few pieces in literature to call out that I think are meaningful applications and showcase some of the diversity in their use cases.

- Convolutional variational autoencoders applied to transient incompressible flows, such that the computational cost can be reduced to make optimization more feasible (as demonstrated here on a 2D Karman Vortex street flow and a 3D incompressible and unsteady flow in an aeronautical injection system) [17].
- A method for learning compact and near-orthogonal ROMs using a combination of a β-VAE and a transformer, tested on numerical data from a 2D viscous flow in both periodic and chaotic regimes. The β-VAE is trained to learn a compact latent representation of the flow velocity, and the transformer is trained to predict the temporal dynamics in latent space [18].
- A method to generate CAD, including the user operations to create the CAD as well, which is based on the transformer due to popularity in natural language processing [19].
- BlastNET (Bearable Large Accessible Scientific Training Network-of-Datasets), which provides machine learnings operable on very high-fidelity (large) datasets in combustion and turbulent flows (as well as a web platform to provide datasets also) [20]. The model put forth is a (3D) convolutional autoencoder, which is comprised of convolution layers and residual blocks.

GNNS

Graph Neural Networks (GNNs) are a type of neural network designed to operate and learn directly on graph-structured data (in our case, meshes from our simulations can be directly input). You can think of them as 'a graph goes in' and 'a graph comes out', with the specific core pieces doing the 'learning' are based on neural networks. In the end, predictions can be made at different 'levels' (graph nodes, or edges). For simulations, the concept of graph representations is fundamental to the usage of graph neural networks, which in essence, a graph is a collection of edges between nodes that parallels the relationships between different cells in our simulation domain. In our case of CAE applications, we would probably be interested in regression most of the time, which means predicting a continuous value at the node-level (and also maybe global 0D metrics). GNNs have gained popularity for CAE simulations in part due to their ability to work on unstructured meshes, which is a great quality since many industrial simulations are forced to use unstructured meshes, since they do not take as long as to setup as structured grids and can capture the naturally complex/irregular geometries.

For data originating from simulations, the discretization of the domain into a mesh can be represented as a graph, which would have scalar results present on the nodes, and then used by GNNs. I really like showing this image for a conceptual starting point from this webpage [21], which represents an image as a rectangular grid of pixels (with RGB channels as arrays). This representation of an image (a smiley face) can equate every pixel as a node that is connected to neighboring pixels/node via edges. Values are stored at each node, which in this case is the RGB values at each pixel (three separate values). The left portion of the image conveys the content of the image, whereas the middle figure is the adjacency matrix. The adjacency matrix helps capture which nodes are actually connected to which other nodes, visualized in a square matrix.

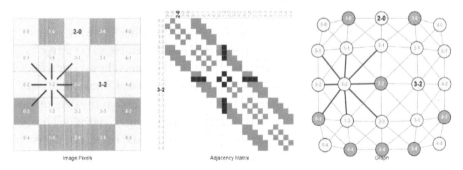

Figure 25: Graph representations of an image (left-most image).

To store the information of the nodes {node 1, node 2, ..., node n} and the edges which connect nodes {(node 1, node 2), (node 2, node 3), ...}, the graph ends up being a tuple. The nodes and edges may have attributes (e.g. directional dependence) which we want to represent for matrix operations (neural network), which ushers in plenty of other variations you can read about in literature. We won't go further into surveying these or types of GNNs, and there certainly are tons of variations, but you could start with reading about: Recurrent GNNs, Graph Convolutional Networks, and Graph Auto-Encoder Networks, to name a few.

In commercial CFD projects, usage of GNNs results in a model that can make predictions of the local results (e.g. pressure values on each node in the mesh in the computational domain) as well as (0D) global scalars for the design (e.g. a single pressure drop value for the entire domain). These models can make inferences on stl files for new candidate designs and are easily used for transfer learning within a discipline by taking a pre-trained model and updating the model on subsequent unseen datasets. The fact that GNNs can work with inconsistent input data, in this case unstructured meshes that are all unique from one simulation to another, make it a suitable choice for industrial simulations that have complicated geometries and are forced to use unstructured meshing styles.

For large scale graphs, neighborhood sampling techniques are used to handle so many points, which results in more efficient training on these large CFD meshes that are many thousands of points. The hyperparameter that controls this passing of messages, or the distance information travels to neighboring points, goes by different names for different models (for example sampling size), but essentially controls how many neighboring nodes to take information from (aggregate) when training. This is crucial, since you can imagine it is not feasible to perform these operations for all neighboring nodes in a highly refined large CFD mesh.

Let's get you a sandbox to play in with GNNs so that you can impose different hyperparameter value sets into your own experiments. While there are several ways to get setup and run this project, one I can recommend is to get a free Google Colab account (includes and a GPU to use for limited hours per week). I offer this clean and good work as a project you can download from the following GitHub code [39] as a start. The only missing piece is the dataset files, which the author says they will provide if you contact them, or download it on their other GitHub project they reference [41]. If you do the latter, you'll have to convert the image files to .csv files (with a binary encoding for wall cells or fluid cells). In any case, I recommend reaching out to the author, and once you have the files you will add these csv files to your 'dataset' folder that's already created in the Github folder from the original repo [39].

The referenced project is for a laminar flow, whereby you provide a random object shape and the model produces a prediction of the pressure and velocity field using graph neural networks [40]. The inputs to the model are the locations of the cells which are labeled as solid, and the output is the pressure and velocity values on all the other cells (the flow-field). You can see a sample result of the predictions versus ground truth from the cfl-minds group [39].

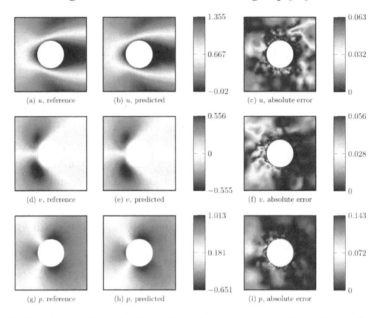

Figure 26: Sample result from the cfl-minds laminar flow project with graph neural networks [39]

For your experiments, you can tweak the network hyperparameters and architecture in the 'params.py' and 'network_utils.py' files. After getting acquainted with replicating the results and running their codes, I think it would be interesting to introduce specific classes of shapes (e.g. a cross-section of different automobiles) and re-train the models on-top of these initial/baseline models trained for random shapes. You could try to see how general your trained models could perform (e.g. expanding to cars, sedans, SUVs, and other classes of automobiles). Additionally, I think you could experiment with increasingly large meshes to see how you could tune your hyperparameters to scale with more datapoints. This is one of the strengths of GNNs which make them a suitable choice for industry compared to other methods which cannot predict over such non-uniform and dense meshes from industrial simulations.

RNNS

A Recurrent Neural Network (RNN) is a type of artificial neural network designed to recognize patterns in sequences of data, such as strings of text, genomes, or numerical time series data emanating from sensors. Unlike feedforward neural networks, RNNs have a unique feature: they have a memory that captures information about what has been calculated so far. In essence, an RNN has internal loops that allow information to persist in the network.

There are many variations of RNNs to accommodate different problems, including the famous Long Short-Term Memory networks, which are a notable improvement for data where long patterns in time are important. LSTMs typically do better than simple RNNs due to having a more suitable architecture to handle the vanishing gradients challenge (updating the weights and biases in the neural network becomes ever challenging with more and more layers due to smaller values of gradients). These are discussed all over the internet and in books, so I will quickly proceed to talk about a few examples. A really interesting thing to see is how different types of architectures can be combined to increase the capability of the model for certain applications. Like for example a convolutional neural network and an RNN.

Hybrid neural network is designed for unsteady flow prediction for vortex shedding over a cylinder [22].

The combination of LSTM with Convolutional Neural Networks (CNN) has been proposed to predict unsteady flows. In this hybrid architecture, the LSTM network is utilized for predicting lift coefficients at each time step, while the CNN aids in predicting the velocity and pressure fields based on the LSTM's predictions.

A U-Net-LSTM framework for rapid and high precision hydrodynamics for a marine application; made of a deep U-shaped network, two LSTM layers, and a skip connection [23].

Compared with the traditional hybrid convolutional neural network-LSTM (CNN-LSTM) framework, the mean square error and mean absolute error are reduced by almost one order of magnitude and two orders of magnitude, respectively, showing that the proposed framework is highly competitive.

Unsteady aerodynamic reduced order model, which is LSTM-based, used to predict the buffet phenomenon [24]

This phenomenon occurs at high subsonic Mach number at high angle of attack, and the model was able to accurately predict force and moment integral quantities and capture both self-sustained unsteadiness and external excitation behavior. Interestingly, for the timeless question of 'how does the model predict

110

outside it's training range', the trends are reasonably captured outside the training range.

Expanding the scope to include physical test data, RNNs and variants have a lot of usage concerning physical measurement data to detect anomalies in the behavior of the equipment, measurement devices, or immediate environment [25]. Some specific examples could be vibration in aircraft engines [26] or active control of wind turbine pitch angle via estimating and forecasting wind patterns [27], among a myriad of others.

TREE-BASED MODELS

Tree-based machine learning models are among the most popular and versatile algorithms in the data science toolkit, prized for their interpretability, predictive power, and ability to capture non-linear relationships. They generally work well on tabular data, which is really common in CAE (and elsewhere!). They are suitable for big, and even huge, datasets where other models may struggle.

At the foundation lies the Decision Tree, a flowchart-like structure where internal nodes represent decisions based on feature values, leading to terminal leaves (think of a tree leaf) that signify predictions. While individual decision trees are intuitive and easy to visualize, they can sometimes be overfit to training data. To combat this, we have Random Forest, which ensemble multiple decision trees, each trained on a random subset of the data, to produce more robust and generalized predictions. Building on the ensemble idea, XGBoost (Extreme Gradient Boosting) takes a gradient boosting approach, iteratively adding new trees that correct the errors of the existing ensemble. This method has gained immense popularity due to its efficiency and performance, often dominating machine learning competitions and real-world applications.

Plenty of thought-provoking notions on how tree-based methods can outperform deep learning methods on tabular datasets can be read in this paper by Grinsztajn et al. [28]. They test highly successful tree based models (random forest, gradient based trees, and XGBoost) versus neural network based models (MLP, Resnet, and two types of transformers) on a variety of tabular datasets to abstract characteristics in the performance of each. Specifically, the author focuses on certain challenges for neural networks. Firstly, analyzing the difference in performance when there are features included that are not useful/significant in finding a correlation. In tabular datasets there are more features included than just those which are essential, which makes sense as 'to be safe' a model builder could include many of the features available as a starting point if they don't have a strong intuition on which could be eliminated. In this paper, the author illustrates performance increases for the neural network-based models as these less useful features are removed, whereas for the tree-based models they are accurate inherently without any manipulation due to their hierarchical structure (which can isolate uninformative features at lower, less influential, splits of the tree structure to ensure limited impact).

To summarize the other main findings from the paper, the next major observation was that neural networks are biased towards overly smooth solutions. Via manipulating the data in the experiment, their observation was that neural networks struggle to fit irregularities, and prefer low frequency functions, compared to tree-based methods, since (for example) decision tree-based models learn piece-wise functions.

Firstly, let's do a topical discussion on random forests. Random forests are made of numerous decision trees but have the advantage of being created on various subsets of the data, such that the final output is based on aggregate average of individual subsets. That helps prevent overfitting, which is a prevalent struggle when using decision trees directly. One other attraction of random forests is their easy-to-use setup (see very simple code below), as well as their ability to be quite accurate on datasets with many input features.

```
from sklearn.ensemble import RandomForestRegressor

# Create regressor
regressor = RandomForestRegressor(n_estimators=100,
random_state=41)

#Fit model for (X,y) dataset
regressor.fit(X_train, y_train)

#Test the model on output data
Y_pred = regressor.predict(X_test))
```

Another inherent benefit in using random forests is their ability to easily gauge the importance 'ranking' of the different features used, as well as a variety of interpretations to the model predictions and inner workings. I will take this opportunity to review some of my favorite plots to make with random forests to analyze and digest the model predictions, since indeed it is quite succinct to setup the model with very few lines of code. I will use a public dataset collected from the 2018 FIFA World Cup [29]. The dataset is comprised of statistics from all the matches in the tournament, whereby the goal of the machine learning model would be to make an accurate prediction on who the man of the match award would be given to. It will be useful to showcase some of the analysis plots/approaches available when using random forest models, so you could be better acquainted with them in your own projects.

Feature importance. Shows the importance of each feature in making accurate predictions, relative to the others in a trained model. This helps in understanding which features are driving the model's predictions. It is a useful start for identifying important features, interpreting the results and tendencies of the model, and providing aid in feature engineering exercises. This simplified code is all you need to begin plotting the importances.

```
# Extract feature importance and plot
importances = tree_model.feature_importances_
plt.barh(feature_names, importances)
```

Output:

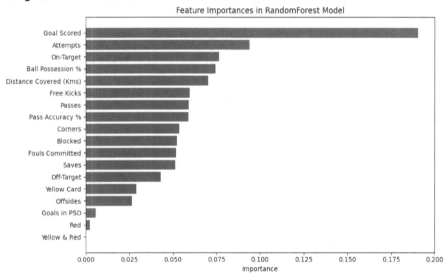

Feature Importances in RandomForest Model

Partial dependence plots. Visualize the way changing one feature affects the predicted outcome by keeping other features constant. It's useful to see how the predictions will vary by changing one input over a broad range in a systematic way where other changes are limited. When using this tool, you would select a feature, define a grid of values to test for that feature, make predictions with the model, and then plot the results: the x-axis is the grid (feature) values chosen and the y-axis will be the prediction. The y-axis is a change in model prediction, relative to leftmost value. Plotted below we can see three of these plots; the left two originating from a decision tree and the right-most from a random forest model. From the goal scored figure, we can see that if a player scores their first goal it significantly boosts their chances to win the award. For the distance covered, we can see a pretty unrealistic result from decision tree classifier model and much less piece-wise (which is probably more realistic) relationship from the random forest model for the same dependence on the distance covered.

```
# Plot goals scored dependence from decision tree model
from sklearn.inspection import PartialDependenceDisplay

Display1 =
PartialDependenceDisplay.from_estimator(tree_model, val_X,
['Goal Scored'])
plt.show()


# Plot goals scored dependence from random forest model
Display2 =
PartialDependenceDisplay.from_estimator(rf_model, val_X,
['Goal Scored'])
plt.show()
```

Output:

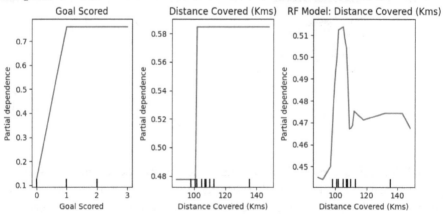

```
*Edit: units should not be in Kms
```

Tree Visualizations. Even though random forests consist of many trees, visualizing a few individual trees can help in understanding the decision-making process. These hierarchical representations are easy to interpret when striving to better understand the decision-making process. Since random forests can easily consist of 100s of trees, looking at one specific tree is of limited value and it's best to have some key things to inspect over a small selection of trees you may survey: analyze the splits across several trees (many trees in the forest splitting on a particular feature value could indicate a strong pattern), evaluate the complexity in the captured patterns but depth of the trees, and as a final example look at trees on your validation set that have particular good or bad performance to understand your model behavior better. When looking at the tree diagram below in the output, here are a few comments on the text included: the splitting criteria is shown at the top and the bracketed values at the bottom indicate the count which go to either 'True' or 'False'. You can directly look at the model hyperparameters to moderate the tree structure/appearance, especially with the minimum count required to split a node, the depth of the tree, etc.

```
import graphviz
tree_graph = tree.export_graphviz(tree_model,
out_file=None, feature_names=feature_names)

graphviz.Source(tree_graph)
```

Output:

Out-of-Bag (OOB) Error Rate Plot: If you use the OOB samples to estimate error, you can plot the OOB error rate as more trees are added. This helps in understanding if adding more trees improves the model and can help conserve compute resources by not adding more trees than necessary when error has plateaued. Training more trees in a random forest requires more CPU time, as each tree is trained independently, so it can be a useful tool to employ early in your experimentation. Below is the relevant code snippet to generate such a plot/analysis once the data is imported and prepared.

```python
# Define the range of trees to evaluate
n_trees_range = range(1, 201)

for n_trees in n_trees_range:
    # Train a random forest with OOB score enabled
    rf = RandomForestClassifier(n_estimators=n_trees,
oob_score=True, bootstrap=True, random_state=42)
    rf.fit(train_X, train_y)

    # Compute the OOB error rate
    oob_error = 1 - rf.oob_score_
    oob_error_rates.append(oob_error)
```

Output:

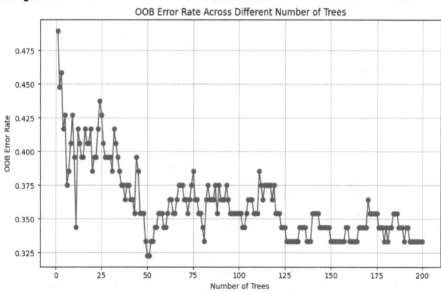

As an example, for random forests applied to CAE, we can cite/discuss a seminal work by Ling [30]. After evaluating several machine learning models, they concluded that random forests were superior for this problem of modeling turbulence in several canonical CFD problems (jet-in-crossflow, flow over simple obstacles, etc.). It's likely this conclusion may get outdated with the advent of more sophisticated techniques to handle large 3D datasets, but from the era of 2015-2020 random forests did among the best at discerning characteristics of turbulence throughout the computational domain (two-dimensional and three-dimensional position). After some careful thought on the input features, like for example constructing a Reynolds number definition based on turbulent scales and wall distance, the hierarchical nature of the random forests was robust enough to correlate the different aspects of the turbulence to different regimes (which makes sense since we know turbulent flow is highly dependent on wall distance). This is a great paper to review in order to get a grasp for the regressive power of random forests in CAE applications.

PHYSICS-INFORMED NEURAL NETWORKS

Physics-Informed Neural Networks (PINNs) have taken the scientific machine learning community by storm since they were first published around 2017. There are dozens of variations that exist in literature for our realm of engineering simulation, many of which have made impressive strides to improve their suitable for use in industrial settings. They are somewhat of a controversy in many circles, as they are very compelling in theory and also for some specific use cases, however, it is at least my opinion that they have not been adopted at a wide scale yet in industry due to some limitations. I would like to provide a brief general overview of PINNs and then conclude with some 'key considerations' (e.g. ongoing challenges) and resources for the readers to benefit from.

First, kudos and a reference to Ben Moseley's blog [42]; PINNs have been explained many publications and articles, but these visual aids are excellent and I would like to use them throughout my introduction to PINNs herein.

The main motivation behind PINNs begins with the ambition to put our physics, or the equations which describe the physics for the problem we are trying to solve (e.g. Navier-Stokes for CFD simulation), into the machine learning model (i.e. neural network) we are trying to use. Alternatively, we can simply use data alone to 'naively' train our neural network, but wouldn't it be beneficial to include some of the mathematics as a help to steer it in the right direction?

For the example on Ben Moseley's blog [42], we can consider a 1D damped harmonic oscillator as shown in Figure 27. Specifically, the position over time of the oscillator which is attached to a spring after it is released. If we take snapshots of the position over time to use as training data, but only for the beginning (early time) instances in the full process, we can train a neural network directly on said data. Of course, the model would fit well for the (orange points) data points used for training, but the predictions outside the window for which training samples were accrued would be really awful (even if you trained for many, many epochs).

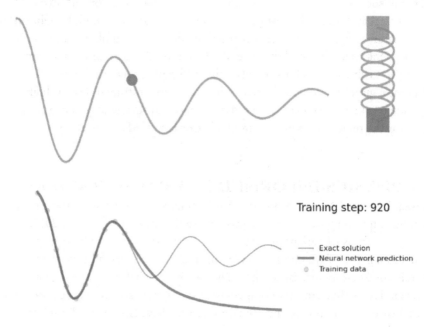

Training step: 920

——— Exact solution
■■■ Neural network prediction
◦   Training data

Figure 27: Top: Position over time for the 1D damped harmonic oscillator. Bottom: Example of a neural network fitting a model to discrete points in the process early-on [42].

To forget about PINNs for a moment, neural networks seek to tune the parameters (weights, biases) to minimize some error that describes the discrepancy between the sample data (u_{True}) and the neural network predictions (u_{NN}). The best error formulation to use is different from one project to another, as it should be specific to the machine learning problem at hand. For simplicity, we can use the mean squared error metric for this problem, as shown below. From this, it's no surprise the model optimizes its weight to match these discrete points and does not care about the underlying physics, or another or points.

$$M.S.E. = \frac{1}{N}\sum_{i}^{N}(u_{NN} - u_{True})^2$$

Now, with PINNs, we modify this loss term from the typical neural network usage in a data-driven fashion; we add another term that is based on the underlying differential equation. With this, we can expect that as the new full error term approaches zero, the neural network model prediction will exactly match/satisfy the ground truth. You can think of the added component in the error definition as the mismatch between the differential equation solution and the neural network prediction, hence the prediction more closely matches as the error term goes to zero. For this problem, we know the equation is:

$$mass \cdot \frac{d^2u}{dx^2} + friction \cdot \frac{du}{dx} + spring\_constant \cdot u = 0$$

As such, our next error term for a PINNs model would be two terms: our original MSE term plus the new one:

$$\frac{1}{N}\sum_{i}^{N}(u_{NN} - u_{True})^2 + \frac{1}{M}\sum_{j}^{M}([m\frac{d^2}{dx^2} + \mu\frac{d}{dx} + k]\,u_{NN})^2$$

The first term is simple enough; for the training data, the loss term is proportionate to the difference in predicted versus ground truth. The second term is trickier, so let's explain it a bit more.

We want to compute various derivatives of the PINN with respect to its input, and then combine them together so that those derivatives evaluate this residual of the underlying differential equation. So if the PINN knew the solution exactly, the second term in the error definition would be zero, but if it's not then the training process can iteratively change the weights and biases so that it may approach zero.

So what you may wonder is what new data we need for PINNs to calculate this error? We have observable samples, which are used in the first part of the error term, but for the second we need to provide more data to evaluate the PINN inside our computational domain. In the context of Moseley's 1D oscillator, we can impose green dots in figure x to calculate the PINNs error at these moments. Notice we have discrete observation points in orange still that were used for training, but we also have now a uniform grid of green points that are used in the PINNs error term calculation that extends over the full range of the problem and far beyond the training data points in orange.

Figure 28: The PINNs model prediction after leveraging points in the domain (green points) for the PINNs error term.

This approach is powerful, and you probably have gained some excitement after reading through this brief example. This is for good reason, as PINNs do have several notable positives:

- If we run a CFD or FEA simulation, we must discretize our domain with a mesh to run the simulation. Especially for industrial cases where the geometries we want to simulate are quite complex, this part of the process can be time consuming. Since PINNs represent a functional solution, they are mesh free and you can train them without making a discretization.

- PINNs are mostly unsupervised, especially compared to other data-driven approaches where you may have data-hungry models that are expensive to generate.

- They work well in 'mixed' regimes and with noisy data.

However, there are some drawbacks that remain as future areas for improvement, which I see as important focus areas before more uptake in industry is possible.

- Most importantly, they often have prohibitive large training times that would discourage a practitioner to implement a PINNs approach, rather than just simulate the points they are interested in. In fact, I have regularly seen PINNs take more than two orders of magnitude more time to train than to do a single simulation. In my opinion, this is their largest barrier for entry, and I would focus on making them faster if I wanted to pursue research in PINNs. Some of the original authors have indeed published improvements over the years, which is very exciting!

- Very expensive to apply PINNs to larger domains with more complicated physics (poor scaling).

Note that PINNs can be used for many scientific purposes and the present writeup is just a focus on PINNs for simulation practitioners. You can also mix physical measurement with simulation in PINNs projects, which of course is another strong incentive to use them.

7.   TUNING YOUR MODELS

Once you have a handful of models you think are suitable to model your dataset, you can and should explore fine tuning of the hyperparameters and other 'flavors' of your model which are possible. For added clarity; by 'have' I mean either you know from experience/intuition there are a few specific models which are well suited for your problem, or that you did some baseline tests with a wider variety of models and have down selected a couple sensible choices to refine.

Let's start with hyperparameters, not to be confused with other 'parameters' involved: a hyperparameter is a value that guides the learning process, whereas the values of other parameters are determined (as an outcome) through training a model. For a neural network, parameters would refer to weights and biases, and for other models the parameters would be different and specific to that architecture. Hyperparameters, as an example for a neural network, would be things like the number of neurons per layer, the number of layers, and the batch size, to name a few. Another example would be number of trees for a random forest model, and number of neighbors for K-nearest neighbor.

It would not be within scope of this book to review the hyperparameters and parameters from various models, so instead let's dive into a code example that conveys how to execute hyperparameter fine tuning approaches. This will work with the famous sonar dataset [31] and showcase three methods: grid-search, random search, and Bayesian optimization. This example will be done on a CPU to ensure less people are alienated from not having GPU access. For background, the dataset is comprised of 208 rows (samples) and 60 columns (inputs). The model trained on this data should be able to judge if the sonar signal is bouncing off a metal object or a rock-like object, and thus is a classification problem.

Firstly, let's import the data and perform some basic handling: load it via the website link which has the data present, split into two Numpy arrays based on inputs and outputs, and then divide for training and testing.

```
url =
'https://raw.githubusercontent.com/jbrownlee/Datasets/maste
r/sonar.csv'
dataframe = read_csv(url, header=None)
# split into input and output elements
data = dataframe.values
X, y = data[:, :-1], data[:, -1]
```

```
print(X.shape, y.shape)
X_train, X_test, y_train, y_test = train_test_split(X, y,
test_size=0.25, random_state=42)
```

Next, we can setup the grid search hyperparameters to use in the search process (with print(type(param_grid)) which outputs <class 'dict'>). Since we will use a random forest model, you'll notice the hyperparameters are specific to such model and things like 'layers' and 'neurons' are not showing up, as the case would be for a neural network. Briefly, in order, these hyperparameters reflect: the number of trees, the maximum depth of each tree, the function by which to calculate how many features to use when building different parts of the tree, and a Boolean designation for whether to use bootstrap samples.

```
# Define the hyperparameters and their possible values
param_grid = {
    'n_estimators': [10, 50, 100, 150, 200],
    'max_depth': [None, 10, 20, 30, 40, 50],
    'max_features': ['auto', 'sqrt', 'log2'],
    'bootstrap': [True, False]
}
```

Next, we can initialize our model, which means that we are creating an instance of the Random Forest algorithm, which is ready to be trained on a dataset. We will use the 'GridSearchCV' class in the scikit-learn library that performs an exhaustive search over our specified parameter values. At last, we will use 'fit' to train the underlying model (rf) on the training data for each combination of hyperparameters (selecting the combination that gives the best performance based on our specified scoring method). Though I have omitted some minor code with print statements, I will include the output text which reveals about a 7% increase in accuracy thanks to grid search.

```
# Initialize a random forest classifier
rf = RandomForestClassifier()
```

```
# Use grid search with 5-fold cross-validation
grid_search = GridSearchCV(rf, param_grid, cv=5,
scoring='accuracy')
grid_search.fit(X_train, y_train)
```

```
# Evaluate the best model from grid search on the test set
y_pred_baseline = rf.fit(X_train, y_train).predict(X_test)
baseline_accuracy = accuracy_score(y_test, y_pred_baseline)
y_pred_grid_search =
grid_search.best_estimator_.predict(X_test)
grid_search_accuracy = accuracy_score(y_test,
y_pred_grid_search)
improvement_grid_search = (grid_search_accuracy -
baseline_accuracy) / baseline_accuracy * 100
```

Output:
```
Baseline Accuracy: 0.8077
Accuracy after Grid Search: 0.8654
Improvement in Accuracy using Grid Search: 7.14%
```

As with many situations, results from a 'black box' are of limited value due to the inability to discern fundamental things like why the results are the way they are, which components (variables) are the most influential on the results, and what the effort-to-reward ratio is, to name a few. With a few simple plots, we can get more insight into what's happening during the grid search operation. Specifically, we can plot the accuracy against two 'knobs' (hyperparameters), shown as separate plots. Information like this is not mind-blowing, but it does provide helpful expectations when deciding in the trade-off between computational cost (via expensive hyperparameters) and accuracy.

```
# Extract mean test scores
mean_test_scores =
grid_search.cv_results_['mean_test_score']

# Extract values of the hyperparameters for all
combinations
param_n_estimators = [param['n_estimators'] for param in
grid_search.cv_results_['params']]
param_max_depth = [param['max_depth'] for param in
grid_search.cv_results_['params']]
```

Output:

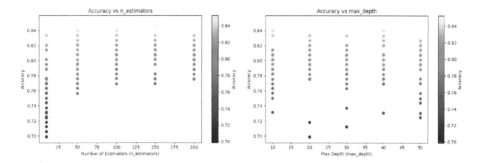

This is a natural segway to the next code snippet example, whereby we want to try random search. From the above two plots, the information is useful however not definitive on exactly which hyperparameters we should pick (since there is a wide scatter in each). Therefore, maybe an optimization approach would be best to iteratively converge on the ideal hyperparameter values. Prior to optimization though, just for illustration purposes, we can first try a random search technique.

This will code snippet will be similar to the grid search approach, but we would like to define the hyperparameter ranges with random distributions. As you may suspect, there is one term n_iter that assigns a number for allocated iterations to perform in the search; specifically, this is the number of combinations of hyperparameters to try. As in previous examples, we will use k-fold cross-validation (the dataset will be split into 5 parts/folds, whereby the data is rotated in the 4 training and 1 validation fold). The result of this exercise is that the best model accuracy is in the same ballpark as the grid-search, which is not very surprising.

```
param_dist = {
    'n_estimators': randint(10, 200),
    'max_depth': [None] + list(randint(10, 50).rvs(5)),
    'max_features': ['auto', 'sqrt', 'log2'],
    'bootstrap': [True, False]
}

# Initialize RandomizedSearchCV with a fixed number of
iterations
random_search = RandomizedSearchCV(rf,
param_distributions=param_dist, n_iter=100, cv=5,
scoring='accuracy', random_state=42)
random_search.fit(X_train, y_train)
```

```
# Obtain the best accuracy
best_accuracy = random_search.best_score_
print(f"Best Accuracy from Random Search:
{best_accuracy:.4f}")
print(f"Improvement in Accuracy: {100 * (best_accuracy -
baseline_accuracy):.2f}%")
```

Output:
Best Accuracy from Random Search: 0.8401

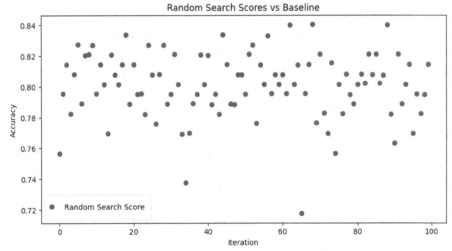

At last, let's try the optimization approach for determining the best hyperparameters. The code snippet below will use a Bayesian optimization approach, which is a probabilistic model that tries to find the minimum or maximum of any function (makes educated guesses where the optimal value might be, based on past data). The model has an initial belief before observing any data (the prior distribution), and after making a guess (or taking in new data) the model updates its belief (which is called the posterior). Bayesian optimization constantly has to decide between exploring new areas, which are regions of the parameter space it's uncertain about, and exploiting known good areas, which are regions where it believes the objective function is optimized.

After importing necessary libraries, we define the function to optimize, rf_cv, which takes in hyperparameters for the random forest classifier and creates the model and evaluates the performance using 5-fold cross-validation. As in the comments, the BayesianOptimization object is initialized with 20 initial random points and then performs 50 iterations of the optimization. After optimization, the best hyperparameters are extracted from the 'optimizer' object and used to train a new random forest classifier ('best_rf'). The best model is then evaluated in 5-fold cross-validation and the output is printed for the user to see.

```python
from bayes_opt import BayesianOptimization
from sklearn.ensemble import RandomForestClassifier
from sklearn.model_selection import cross_val_score

# Define function to optimize
def rf_cv(n_estimators, max_depth, max_features):
    estimator = RandomForestClassifier(
        n_estimators=int(n_estimators),
        max_depth=int(max_depth),
        max_features=min(max_features, 0.999),
        random_state=42,
        n_jobs=-1
    )
    cval = cross_val_score(estimator, X_train, y_train,
scoring='accuracy', cv=5)
    return cval.mean()

# Initialize Bayesian Optimization
params = {
    'n_estimators': (10, 200),
    'max_depth': (1, 50),
    'max_features': (0.1, 0.999)
}

optimizer = BayesianOptimization(f=rf_cv, pbounds=params,
random_state=42)
optimizer.maximize(init_points=20, n_iter=50)

# Extract the best parameters
best_params = optimizer.max['params']

best_rf = RandomForestClassifier(
    n_estimators=int(best_params['n_estimators']),
```

```
    max_depth=int(best_params['max_depth']),
    max_features=min(best_params['max_features'], 0.999),
    random_state=42,
    n_jobs=-1
)

best_rf.fit(X_train, y_train)

# Obtain the best accuracy
best_accuracy = cross_val_score(best_rf, X_train, y_train,
scoring='accuracy', cv=5).mean()
print(f"Best Accuracy after Bayesian Optimization:
{best_accuracy:.4f}")
```

Output:

iter n_esti...	target	bootstrap	max_depth	max_fe...
1 123.7	0.7369	0.3745	47.54	44.19
2 174.6	0.8139	0.156	7.8	4.427
3 194.3	0.8202	0.6011	35.4	2.214
4 44.85	0.8018	0.8324	10.62	11.73
5 65.33	0.7823	0.3042	26.24	26.48
6 200.0	0.7438	1.0	0.0	47.96

...

Hyperparameter tuning is a great tool to have available to improve the more accuracy, but we will discuss some cautionary notes and drawbacks of such as a means to segway into additional recommended steps and tools that can and should also be considered. Don't get too discouraged though; any commercial product worth its salt would have built-in packages that allow you do this easily without much thought.

- Hyperparameter tuning takes a non-trivial amount of time and can be infeasible for projects that have training-time constrains. One aspect of wasted time is that during the process some models may take time to be trained that are obviously poor choices that an experienced practitioner would never consider.

- Hyperparameter tuning can lead the user to increases in model size that can become prohibitively large and infeasible models.

- **Hyperparameter tuning can lead to models that are strongly overfitted to dataset.** This can be especially true if the user opens up a huge hyperparameter space to consider, providing freedom to pose models that are massively overfitted.

Hyperparameter tuning is often described as an art more than a science, inefficient, and highly problem specific. I stumbled across a cool approach, which is one of many possible ways to do it, that prescribed a way to help address this by focusing on batch size and learning rate first. It seeks to find some nearly flat directions in hyperparameter space, as you can see illustrated in the image below in Figure 29. This simplifies the search problem, which can help you optimize your hyperparameters in an efficient direction.

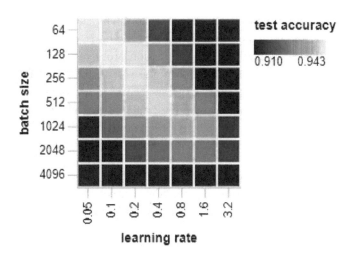

Figure 29: One approach for hyperparameter tuning, whereby you change batch size and learning rate first to note the change in error.

So at this time, let's stop and reflect on some tools we have that can help increase our model accuracy when we are training and evaluating different model variations. This will be at a summary level of detail since many of these concepts we have already discussed in some form.

- Feature engineering. Feature creation: probably using domain specific knowledge, derive new features from existing and then apply techniques (e.g. recursive elimination) to down-select a minimal and manageable effective set. Also, manipulate your features via compression techniques (e.g. PCA) and/or scale your features for better results (e.g. normalization or standardization).

- Regularization. Techniques like Lasso (for L1) and Ridge (for L2), as two examples, can be useful to reduce the overfitting nature of the model by adding penalties in training. This promotes more simple and generalizable models.

- Resampling. Whether it be from class imbalances in classification problems, presence of outliers, non-uniform data density, as a few examples, resampling can be helpful to address issues in the data that might hinder the development of robust and accurate predictive models. Previously, we gave such examples like SMOTE, bootstrapping, and undersampling.

- Data cleaning/transformations. As discussed, these techniques are fundamental to the pre-processing stage in the pipeline but can have big impacts on the model performance. Many methods, tools, and procedures fall into this category. Specific to regression problems, we can name things like adding small amounts of random noise to the data, applying transformations like logarithmic or exponential, domain specific augmentations (e.g. using non-dimensional numbers like Reynolds and Nusselt number instead of velocity or heat transfer coefficient), generating synthetic data with methods like VAEs or GANs, to name a few.

- Hyperparameter tuning. Tweaking the model hyperparameters is often a very effective tool to improve the model accuracy. Normally, I would not opt to do this first thing in a project, as usually there are more universal improvements you can make to your dataset which would benefit many models, not just optimized ones. Tools like Grid Search, Random Search, and Bayesian Optimization can automate this process and speed it up for you significantly.

The above concepts can be overwhelming; there are a lot of things you try which leads to a large number of total combinations possible. It also doesn't help they are fairly technical and may require analysis with direct oversight. Have you heard of AutoML? This is actually your silver bullet; a good AutoML capability should rapidly try all of the following types of changes, in addition to different model architectures. Whether a feature in a commercial product, or your AutoML library of choice, using AutoML can save you a lot of time and should be much more exhaustive than just trying sets of hyperparameter values for a specific model architecture.

Now let's turn our discussion to underserved concepts (not discussed too much in this book so far). First, we can start with the concept of ensembling.

Rather than making predictions from only a single model, we can combine multiple models to produce a single prediction. The motivation is to improve the overall performance and robustness compared to using a single model. Certain challenges exist when using single models to make predictions and can incentivize us to try ensembling: the model is very sensitive to the inputs and possesses high variance, the model has low accuracy over some of the entire data range, the model relies mostly on a small number of the total feature set. I personally have often had success with this for spatial field predictions to address challenges of capturing small (but important) details in the scalar field over a wide range of operating (boundary) conditions.

There are several examples for ensembling, but let's just review a select few, starting with stacking. In stacking, we take individually trained models and draw a subsequent prediction as the result of each (base learner predictions become input to the meta-model for the final prediction). Similar to a first layer, which has all the models, and the second one is for making the final prediction. This is especially useful when you are dealing with models that are unable to predict over the full problem space but perform well over the majority of it.

Let's demonstrate a stacking example on the Iris dataset from a Google resource page I found [32]. Suppose you train a decision tree, a neural network, and a support vector machine (CLF1, CLF2, CLF3) on your data. For a given input, these three models produce three predictions. These predictions are then used as inputs for a logistic regression model, which produces the final prediction. In can be a good idea to use a variety of complex and diverse base models to benefit the final prediction. In this sample case, you see a 0.02 improvement as compared to the single best model.

```
iris = datasets.load_iris()
X, y = iris.data[:, 1:3], iris.target
```

```python
clf1 = KNeighborsClassifier(n_neighbors=1)
clf2 = RandomForestClassifier(random_state=1)
clf3 = GaussianNB()
lr = LogisticRegression()
sclf = StackingClassifier(classifiers=[clf1, clf2, clf3],
meta_classifier=lr)


label = ['KNN', 'Random Forest', 'Naive Bayes', 'Stacking
Classifier']
clf_list = [clf1, clf2, clf3, sclf]

fig = plt.figure(figsize=(10,8))
gs = gridspec.GridSpec(2, 2)
grid = itertools.product([0,1],repeat=2)

clf_cv_mean = []
clf_cv_std = []
for clf, label, grd in zip(clf_list, label, grid):

    scores = cross_val_score(clf, X, y, cv=3,
scoring='accuracy')
    print "Accuracy: %.2f (+/- %.2f) [%s]" %(scores.mean(),
scores.std(), label)
    clf_cv_mean.append(scores.mean())
    clf_cv_std.append(scores.std())

    clf.fit(X, y)
    ax = plt.subplot(gs[grd[0], grd[1]])
    fig = plot_decision_regions(X=X, y=y, clf=clf)
    plt.title(label)

plt.show()
```

Output:
```
Accuracy: 0.91 (+/- 0.01) [KNN]
Accuracy: 0.93 (+/- 0.05) [Random Forest]
Accuracy: 0.92 (+/- 0.03) [Naive Bayes]
```

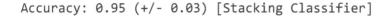

Accuracy: 0.95 (+/- 0.03) [Stacking Classifier]

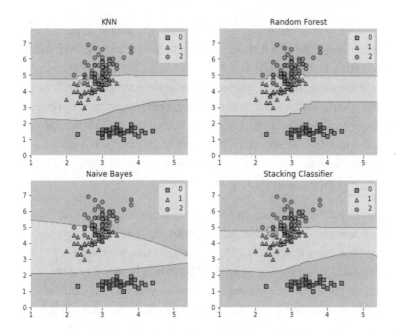

Without code examples, we can summarize a few other types of ensemble methods in list-form:

- **Bagging**. Train multiple models on different subsets of the training data (with replacement) and then aggregate their predictions. It's very similar to bootstrapping, the only difference being it uses a subset of features. Example algorithm: *Random Forest*. An ensemble of decision trees. Each tree is trained on a different subset of the full dataset and often uses a subset of the features as well. The final prediction is made by averaging (for regression) or majority voting (for classification).

- **Boosting**. Train models sequentially, where each new model corrects the errors of its predecessor. Example algorithms: *XGBoost, CatBoost, and Adaboost.*

- **Voting**. Combine predictions from multiple models through a voting mechanism. This is less computationally expensive than stacking, since it's generally just combining the predictions without subsequent modeling. Examples: each model votes for a class and the class with the majority is chosen (hard voting), or probabilities predicted by each model are averaged and the class with the highest probability is chosen (soft voting).

As a departure from ensemble methods, we can talk about another important concept: transfer learning. Transfer learning is a machine learning method where a model developed for a task is reused as the starting point for a model on a second task. For illustration, imagine training a model for predicting the flow in a 2D plane around a bank of cylinders, and then using that model as a starting point for predicting the flow over a bluff body (an automobile). With similar characteristics in the way flow separates from a surface and interacts with incoming flow in the mainstream, this is a reasonable example for getting a head start with a base model rather than starting from scratch.

The benefits extend beyond just more accurate and generalized models, as it can also help with data scarcity, faster convergence (computational efficiency), and continual learning (just a few examples). This is a huge topic to cover for commercial strategy and value for adopting AI/ML methods, as it can help rapidly provide information early-on in design cycles leveraging past datasets from previous work. For example, creating machine learning based models from past generations of vehicles designed and then used for a wide variety of predictions during a current design campaign; while the accuracy may be limited, directional estimates for performance are probably trustworthy. This helps 'shift left' in terms of providing information to the designer earlier in the development process than would ordinally be available. We will depart from commercial and strategic discussion and move to more of a hands-on perspective.

The transfer learning field encompasses several strategies, each with its own nuances which will vary from case to case (e.g. what model you are using). In the most immediate form, transfer learning can be done by simply training one model on a dataset and then using that model as the starting point for a subsequent training exercise on a new dataset. This is called fine tuning, and there are a lot of constituent choices to make in the process. For example, do you let all parts of the model (e.g. layers) change, or do you hold some fixed based on maintaining continuity with the original model? One notable example of this during the ChatGPT we just entered; models like BERT, GPT, and other transformers are often pre-trained on large text corpora and then fine-tuned for specific NLP tasks like sentiment analysis, question-answering, or language translation. Additionally, this is quite common in CNN models as well, whereby a model (e.g. EfficientNet) is originally trained on a vast image dataset, like ImageNet.

Let's provide an example with the following flow:

- Model #1 is trained ahead of time on Dataset #1
- Model #1 and Model #2 are both CNNs with a series of convolution layers, followed by dense layers, and then convolutional layers again. Resembling an encoder-decoder architecture
- Load Model #1 weights from a file
- Transfer weights from Model #1 to Model #2
- Set the weights to 'non-trainable' for the last n (e.g. 3) number of layers

```
#credit:
https://github.com/envfluids/2DDDP/blob/main/DDP_CNN-TL.py


model2.load_weights('./weights_cnn') # load the weights of
the previous model
for layer in model2.layers[2:-3]:
    layer.trainable = False#
```

- Then, Train Model #2 on 'Dataset #2. This code snippet trains the Model #2 on Dataset #2 whereby the training samples are up until sample index number N, and validation uses sample N until the end of the array.

```
History = Model2.fit(input_datatset2[0:N,:,:,:],
output_dataset2[0:N,:,:,:],
        batch_size = batch_size, shuffle='True', verbose=1,
epochs = num_epochs,

validation_data=(input_dataset2[N:,:,:,:],output_dataset2[N
:,:,:,:]))
```

- Lastly, make predictions with Model #2

EVALUATION METRICS

What are evaluation metrics? They are quantitative tools to assess a machine learning model performance. They are crucial for understanding how well a model is performing, for comparing different models, and for guiding the model refinement process.

When tuning and tweaking your models, the choice of which metrics to use in your model training and inference is significant. Without a basic understanding of the metrics you are using, your efforts could be ineffective. The metrics you pick need to be appropriate for your circumstance and picking the right ones can shed light on why the models predict what they predict. This portion of the book is covering information widely available online, so I will be succinct.

Let's start with a practical CAE example to illustrate why our choice of metric is important. Let's say we have generated a dataset comprised of FEA simulations on a turbine blade to train a machine learning model and we are ready to pick which metrics we want to use to evaluate how useful the trained machine learning model is for our project. In this case, the simulation and machine learning model are used to predict a 3D field of stresses, while caring the most about accurate predictions of high stress areas. So, what approach do you take to formulate the appropriate evaluation metrics? Do you take the single highest stress value and see if the model can predict it, or perhaps the average error present in the predictions of the highest 1% of cells? Additionally, do you use absolute units when describing the error in the model or do you make things normalized (e.g. percentages)? These are all important questions, and the answer will be a combination of implementing business logic (goals for that specific project on the turbine blade) with sound machine learning reasoning (in this case, picking the right evaluation metric).

By the way, that was a real project I completed, and my decision was to define an acceptable error band, in absolute units (MAE), and then to count what percentage of cells exceeded that acceptable range within the entire dataset partition used for evaluating the model after training was completed. Further, that subset of cells with high error can be compared to the subset of cells which high had stress values. From this, you can train several models and then decipher which versions are best at predicting high stress areas the most accurately.

Let's kick-off the discussion now on evaluation metrics for regression models. The primary goal of regression is to predict a continuous target variable (based on one or more input variables). Regression models produce a quantity. For instance, predicting the price of a house based on its features, predicting a person's weight based on their height, or forecasting sales for the next quarter are regression problems.

RMSE. Root mean square error (RMSE) can give the practitioner an idea on how much error to expect from the model, especially since RMSE provides error metrics in the same unit as the target variable, making it more interpretable than mean squared error and other metrics. RMSE is one of the most common and preferred metrics, but in itself is incomplete to provide a full characterization on how the model is performing. For example, in datasets which have many outlier

136

points you may want to compliment this metric, or entirely replace it, with the Mean Absolute Error (MAE).

Equation 4: Root mean squared error definition, based on n samples.

$$RSME = \sqrt{\sum_{i=1}^{n} \frac{(y_i - \hat{y}_i)^2}{n}} = \sqrt{Mean\ Squared\ Error}$$

In machine learning discussions you could hear the phrase 'Euclidean norm' (also called L2 norm), which is worth mentioning here since it is of similar form to the RSME. The term "norm" refers to a function that assigns a positive length or size to each vector in a space. Since the equation for such is also a square root of the sum of squares, there is a resemblance to RMSE in that regard. Essentially, if you think of the residuals (the differences between observed and predicted values) as components of a vector, then the RMSE gives the length (or magnitude) of that vector in the Euclidean space, which is exactly the L2 norm of that vector. However, RMSE specifically applies this process to the errors between predicted and actual values in a dataset, while L2 norm can apply to any vector, representing its "length" in the n-dimensional space.

As the index for the norm gets larger, so does its sensitivity to outliers of large value [33]. This is precisely why well-behaved datasets (gaussian) do well with RMSE as the metric to use, but in the case where extreme values are present RMSE can be more sensitive than other metrics (e.g. MAE).

MAE. MAE measures the average absolute error between the true values and the predicted values. Like RMSE, it provides a direct interpretation in terms of the average absolute error made by the model since it is in the same units.

Equation 5: Mean absolute error definition.

$$MAE = \frac{1}{n}\sum_{i=1}^{n} |y_i - \hat{y}_i|$$

You can see that the MAE calculates the average of the absolute differences (residuals) between observed values and predicted values. If you were to scale this by the number of observations, it would effectively be computing the L_1 norm of the residuals vector (which computes the "distance" as the sum of the absolute values of its components, resembling MAE). Hence, MAE is related to the L_1 norm, also called the Manhattan norm, of the residuals vector.

The L_1 norm, commonly referred to as the Manhattan norm (or Taxi norm or City block distance), is named after the grid layout of streets in Manhattan, New

York. If you imagine trying to travel from one point to another in Manhattan, because of the grid structure, you can't just travel "as the crow flies" diagonally across blocks. Instead, you'd travel in vertical and horizontal straight lines that resemble the movements on a grid; you move a certain distance in the north/south direction and then a certain distance in the east/west direction (or vice versa. When calculating the L1 norm of a vector, you sum the absolute values of its components, which is analogous to calculating the total distance traveled along the axes of a grid to move from one point to another.

R^2. At its core, R-squared (R^2) is a measure that tells you how well your model's predictions match the actual data. In simpler terms, it gives you a score between 0 and 1, and the closer it is to 1 means the better your model is at making accurate predictions. R^2 represents the proportion of the variance in the dependent variable explained by the independent variables in a regression model.

Adjusted R^2. While R^2 provides an overall measure of the goodness of fit, it doesn't account for the number of predictors in the model. By predictors, we mean the input factors or variables that you believe have an influence on the output you're trying to predict. This is where the adjusted R^2 comes in handy; the adjusted R^2 takes into account the number of predictors in the model, penalizing for the inclusion of irrelevant predictors. If unnecessary predictors are added to a model, the adjusted R^2 will decrease, even if the regular R^2 might increase; which is helpful for us!

These are some of the main ones for regression, but of course there are more you could learn about. Let us proceed to talk about classification metrics.

To set the stage for when you might find yourself reaching for these metrics for use in your classification project, you'll probably hold some samples out of the training exercise and make a series of predictions afterwards from different models on these samples. Then, you'll base your identification of which model is best as the one which correctly predicted the most samples with the correct classification. Maybe you'll even repeat this exercise several more times, on other withheld samples, to see how the picked model would perform. We will cover certain terms herein you can use to assess the different models' performance.

Let's craft a simple example to use in the definitions to help make these terms stick with you.

Sample Project Goal: Make a machine learning model that can classify which individual parts in an underhood vehicle thermal/fluid simulation would surpass safe ranges of temperature during operation (equivalently, identify which parts that fail due to too high of temperature loading). Each part could be saved as an image file, as a simple example. The model predicts '0' for 'safe parts' and '1' for parts that go into unsafe temperature ranges.

Sample: Any data (image file) provided to the model, as input, for training or inference.

Class: A set of discrete output variables that can individually be used to describe the different types of parts in the simulation. It can also go by 'labels', as in 'each part is labeled as a certain class'.

Accuracy. The percentage of samples that the machine learning model correctly classified (labeled) among the full set. If you pass 10 parts to the model for inference and 9 are correctly classified, no matter what the class is, it is 90% accurate by this nomenclature. Note that this metric can be misleading for imbalanced datasets; for example, if 9/10 parts were actually safe ('0'). Even if my 'model' just predicted 'safe' for every part it received no matter what, it would still achieve 90% accuracy, despite not having any fidelity.

Let's expand our vocabulary here. Accuracy can be broken down into parts. A true positive is when the model correctly predicts the 'unsafe' parts, which more generally means the model predicts the positive class correctly. It means the model was able to identify a real instance of the condition it was looking for, like if you made a model to identify spam emails and a given spam email was correctly labeled as spam. A true negative however, means that the model correctly predicts the negative class; emails that were not spam, individual parts that were safe in operation. These terms are useful to describe the modality of the accuracy; for example, is the model only good at predicting positive cases and we didn't realize that because we had such few negative samples?

$$Accuracy = \frac{\sum True\ Negatives + \sum True\ Positives}{Number\ of\ total\ samples}$$

Precision: The percentage of predicted '1's that are actually '1's. So, of all the samples that were classified as unsafe, how many are correct and actually unsafe.

When putting an equation down to compliment the couple sentences here, you can see we need to define another term. A false positive means the model has predicted a positive condition on a sample when in reality none exists. Whether it be a legitimate email that was incorrectly labeled as spam, or a specific part in the vehicle assembly that the model believes will fail due to thermal overloading (when in reality it will operate safely), these are some examples of a false positive.

$$Precision = \frac{\sum True\ Positives}{\sum False\ Positives + \sum True\ Positives}$$

Specificity: The percentage of predicted '0's that are actually '0's. So, of all the samples that were classified as safe, how many are correct and actually safe.

$$Specificity = \frac{\sum True\ Negatives}{\sum True\ Negatives + \sum False\ Positives}$$

Recall: Of all the samples that are unsafe, how many were correctly labeled as 'unsafe' by the model prediction.

$$Recall = \frac{\sum True\ Positives}{\sum True\ Positives + \sum False\ Negatives}$$

At last, we can describe a false negative as a scenario where the model fails to identify a true occurrence of the positive condition. The model would predict that an email is authentic while it is actually spam, or that a part would operate safely in its thermal condition whilst it would indeed actually fail.

We intentionally finish with false negatives to convey that not all model classification errors are treated equally, and it varies from project to project on which metrics you should care about the most. For some critical applications of classification model, where failure is exceedingly horrible to go undetected, you may tailor your machine learning efforts to reduce your false negative rate (even at the expense of higher false positives, for example). Think about cancer screening; I would much rather go through the scare of falsely claiming cancer present from one of my tests, only to realize later after lab tests the classification was wrong, than have cancer go undetected for years and get worse over time. Or a faulty part in an aircraft that goes undetected and then causes a critical failure during flight, whereas the alternative consequence could be a false positive would represent added cost and down-time due to replacing parts that did not need replacing. Understanding and minimizing false negatives is crucial in areas where failing to detect positive instances can have serious consequences.

Well, this is a lot to keep up with and theoretically to track each metric individually for every model prediction in your project. And as always, effective communication (both to stakeholders and team members) is key in successful data science projects. The **Confusion Matrix** is a great way to distill this information down into a succinct form so you can assess each of these errors types from your model predictions.

An illustration of the confusion matrix is shown below. In the matrix, the rows are the ground truth label, while the columns are the machine learning model predicted label (sorry to say this, but sometimes it is flipped and opposite of this…it just varies from person to person). Usually, you would classify some finite number of parts in your project from your model inference and then place a number in each quadrant to indicate the sum of each occurrence for each (how many TPs, FNs, FPs, and TNs). So, with our working example of the underhood assemble of parts example, we would classify for example 100 different parts as '0' or '1' and then sum the number of true positives, false negatives, etc. If you move along the diagonal, you'll be looking at correctly labeled samples. Off-diagonal numbers represent number of samples that are incorrect, which can be various types of error based on their location.

Predicted Labels

		Positive	Negative
True Labels	Positive	True Positive	False Negative
	Negative	False Positive	True Negative

When selecting these metrics, you must consider the specific context and objectives of the classification task, as each metric offers a different lens through which the model's performance can be understood and interpreted. Classification problems will be less frequent than regression for our CAE application, so I feel it is sufficient to move on.

8. DATASETS AND PROJECTS

When meeting strangers for the first time, there's one question I have been asked far more than any other "how can I learn AI and start adding it to my resume?". One juncture in the learning pathway I always refer to is the value of hands-on learning. While there are plenty of projects and tutorials out there for mainstream machine learning, there are of course orders of magnitude less for our niche little world of engineering simulation. I'd like to use this section to point you in the right direction with helpful engineering datasets, projects, and in some cases, a few guiding steps with such. In general, I think these are good introductory projects to do which should lead to things you can put on your resume (with a bit more work). I find that publishing your research work is obviously a great way to have things to talk to recruiters/hiring mangers about, however the cost associated with publishing, traveling, and attending is often high enough to be prohibitive for individuals paying out of pocket. I hope these projects can help you gain experience while leading you to more advanced projects, and perhaps add things to your portfolio website, without these associated costs.

3D ResNet-like autoencoder for turbulence prediction

The Bearable Large Accessible Scientific Training Network-of-Datasets (BLASTNet) [37] provides both reacting and non-reacting datasets for machine learning practitioners. These datasets originate from high-fidelity simulations and have been processed to be very easily accessible by all! They are distributed on Github and Kaggle so you can run the codes with free compute to try these datasets out.

They really democratize some great resources to enable beginners: nearly 5TB of high simulation data in a convenient form, 13,000+ lines of code to streamline training, over one hundred pre-trained weights for physics problems, and knowledge sharing via conferences and seminars.

Let's start with the Kaggle tutorial [38] by Wai Tong Chung. This notebook focuses on (efficiently) handling and implementing a TensorFlow model with multi-GPU support upon a dataset that is high-dimensional and complex. This a regression model for predicting turbulent quantities at a high spatial resolution.

Since the purpose of the book is to essentially demystify the machine learning pipeline, it's good to have a variety of examples like this one to show the steps in a project from start to finish. In this case, there are distinct steps made clear in the code to complete the project. First, in a very typical way, you can see the relevant libraries that are brought to complete the project and the effort needed to ensure you have multiple GPUs allotted for this notebook. Next up, is the handling of metadata. I touched on this in book very briefly, so it's good we can talk about it now. Metadata is important because while the machine learning model does predict a lot of data that is locally distributed over the cells in the domain, we also rely on more data that is associated with the different simulation runs. This could be called global data, metadata, or others, and is usually stored in a json file (or text file if it's extremely simple and small amount of data). In this case, the metadata variables are the number of points (resolution) in different directions, the number of snapshots, the variable names (e.g. P_Pa, for pressure in units Pascal), boundary condition information, among other things. These values will be used throughout the study, so it's important we process/store it correctly.

The next portion of the notebook is good for learning how to manipulate data and pass it from step to step; the author applies a filter that blurs the 3D data so that it can be used as a low-resolution input, and then down-samples the arrays. The data is then labeled appropriately (features X, inputs Y). To do these operations, and similar ones in many other machine learning projects on spatial/image data, you often create helper functions that can perform these manipulations on your data. A small excerpt to copy/paste from the Kaggle notebook [38] is shown here to illustrate how to blur and down-sample the data.

```
#This blurs and downsamples our data
def my_gauss_filter(phi,fw):
    return gaussian_filter(phi,fw)[int(fw/2)::fw,int(fw/2):
:fw,int(fw/2)::fw]
```

Along the way, you will see some steps in the notebook which makes plots and reports for different portions of the dataset. For example, you will see the notebook convert TensorFlow tensors to NumPy arrays to make it easier to visualize the data, because NumPy arrays are a standard format that can be easily plotted and analyzed using various libraries like Matplotlib. Sanity checks like this are really important to ensure you can identify problems incrementally in your pipeline, and also so that you can have a physical grasp of your data and therefore can make effective modeling choices.

However, moving to the next step, you will use TensorFlow Dataset objects. The motivation behind this is to streamline the process of feeding data into the machine learning model for training; this is very common for complex and large datasets in both PyTorch and TensorFlow (with their own respective formats). The Dataset object in TensorFlow helps you efficiently load your data, preprocess it, and ensures the model can be trained smoothly without unnecessary computational overhead. This is particularly relevant when dealing with large datasets or when requiring complex data transformations (which indeed our data in this case is huge).

The notebook then creates a 3D ResNet-like autoencoder, as shown in a diagram in Figure 30 and from a related publication by the Kaggle notebook author [37]. As discussed, CNNs have a better affinity to learn spatial data compared to other models, since they have a sliding filter that can preserve scalar distributions through space. The autoencoder characteristic is key in this architecture; if you are familiar with proper orthogonal decomposition or principal component analysis from more traditional numerical approaches in our mechanical and aerospace engineering toolbelts, then you can think of this as a similar approach but with much more implicit capability to capture non-linearity. This raw spatial scalar data can be learned by the encoder and then passed to the decoder to generate complex predictions. No free lunch theorem applies though, as this is quite an expensive numerical approach to employ, with approximately 1M trainable parameters in the 93 layers of the network. With modern GPUs however, this is no problem.

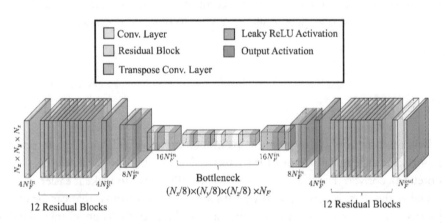

Figure 30: The three-dimensional CNN architecture used.

One of the final stages of the notebook is to define checkpoints for the trainings. We didn't really talk about checkpoints at all so far in the book, so this is a good occasion to do so. Checkpoints are saved states of a model during training. They capture the exact configuration of the model, including the weights for each node, at a specific point in time. They provide us with a handful of conveniences, such as if your computer crashes (loss of power or internet) your model training can revert to a recent saved state, rather than start all the way over from scratch. Additionally, if you are running many models in parallel at once with hopes of identifying the best model to pick, you can see how the accuracy is along-the-way during the training process by pulling the results at certain checkpoints, rather than wait until the entire training process is done for many models at once. As you can imagine, this is an immense resource saver; you can free up space in the queue to run other projects on your GPUs rather than let these jobs run until they are finished training, you will save time, and you will save money if you're paying hourly rates to use certain hardware and machine learning services (e.g. AWS cloud-based machine learning products).

The final steps in the notebook are some of the most fun, in my opinion. You will train the model and then makes some inference/predictions on the data samples that are set aside from evaluation. To me, the most fun aspect is the creativity in how you evaluate the accuracy of the model. You can use metrics we talked about earlier that are 0D in nature (e.g. MAE), and while they are useful for an overview of performance (e.g. pointing out which cases struggled the most during training), you lose a lot of information and insight into the nature of the predictions. Further, you can indeed inspect one-by-one the local predictions for different samples, but at the expense of the pain of doing it manually and looking through so many different raw fields. As such, you can enjoy being creative when crafting insightful and efficient ways to process the predictions. Here are some thoughts which serve as a starting point in your experimentation:

- Make a new variable that is essentially a threshold to filter out cells from the full field of predictions. For example, only 'grab' the cells from each prediction which have the highest turbulence. Then, analyze the accuracy of the model predictions on this subset. This can help us answer things like "is the machine learning model better at prediction highly turbulent regions or lower turbulent regions?".
- Make new and 'physical' 0D metrics. For example, in this case where mixing may be directly related to the turbulence, you could make a script that can calculate circulation, turbulent fluctuations, or other similar things for each case based on the data field. Then, run the calculation of such parameters on both the ground truth data as well as the machine learning predictions. This can give you more insight into what the models can predict well globally, and what they cannot. This can also help you tune hyperparameters if you have a more detailed understanding at where the model is struggling.

- Perform various splitting (e.g. k-fold validations) on the dataset and evaluate the model accuracy on some typical 0D metrics, as well as the ones that you may have created in the above bullet point.

Turbulence Super-Resolution with NVIDIA Modulus [35]

NVIDIA Modulus is a sophisticated open-source AI framework designed to solve complex multi-physics and engineering problems using deep learning. It combines physics-informed modeling with data-driven approaches, enabling high-fidelity simulations and the ability to tackle optimization and inverse problems. If you have heard of it before, you probably mainly think of them for their physics informed neural networks, however they have other capabilities to create high-fidelity, parameterized, surrogate deep learning models. This is a playground for beginners can lead to projects worth including on your resume.

Traditionally, NVIDIA recommends installing Modulus using a container, as that gives you the ease of use and flexibility that is sufficient for most users. The exception being if you want to contribute to Modulus, as an advanced AI researcher, then clone the source code from Github and develop on the fork.

Before jumping into the turbulence super-resolution problem that they provide on their hub as a tutorial, I want to point out an additional way to setup Modulus which might be useful for some people. This is only if you want to jump in as fast as possible to run the tutorial and then move on to other projects, but not recommended for regular use of Modulus. I just want to provide a way to run the tutorial with about 5-10 minutes of setup time, to be as inclusive as possible for the readers. This method is called a 'bare metal install', which installs the software directly on the hardware of a system, without any intervening layers like virtualization or an operating system containerization platform (like Docker). You can do that in Google Colab for example, ensuring you are running on an NVIDIA GPU ('nvidia-smi' command will do the trick to check), with a really rapid setup. Here is some code to guide you.

```
#Install various specific versions of required libraries
that are dependencies for Modulus
!pip3 install matplotlib transforms3d future typing numpy
quadpy numpy-stl==2.16.3 h5py sympy==1.5.1 termcolor
psutilsymengine==0.6.1 numba Cython chaospy torch_optimizer
vtk chaospy termcolor omegaconf hydra-core==1.1.1 einops
timm tensorboard pandas orthopy ndim functorch pint
```

```
#you should run these in separate cells
!pip install hydra-core==1.1.1

#This command sets up Git Large File Storage (LFS), which
is necessary for handling large files in the Modulus
repository
!git lfs install

#This command clones the Modulus repository from NVIDIA's
GitLab. The URL includes authentication credentials
!git clone https://yourname:your-secret-
token@gitlab.com/nvidia/modulus/modulus.git

#navigate to correct folder in directory and run setup file
%cd ./modulus/
!python setup.py install
```

There is one tricky step here, which is for the line "!git clone
https://yourname:your-secret-token@gitlab.com/nvidia/modulus/modulus.git
". It's tricky because the NVIDIA Modulus library is private, so you need to get
access first by setting up an account and logging in (which is free and open).
Inside the NVIDIA 'Modulus GitLab Repository Access' page, you'll need to
make a submission requesting access (you'll need to provide a GitLab username,
so go ahead and make an account there too if you don't have one). One you fill
out the form, you should get an email providing access to the Modulus GitLab
repository. After you are granted access, you'll go into your profile at GitLab and
then click 'access tokens'. You'll want to select the option for 'read_repository',
and then click 'create access token'. You should see a new code appear, and
you'll want to copy that and insert it into the URL you use in the code line that
starts with "!git clone https://yourname:your-secret-token@gitlab.com/nvidia/
modulus.git" between the // and the @. This is the least straightforward step,
so if you make it through this then you're in the clear! You only have to do this
one time.

Next, you want to use git to clone the examples folder, so you can navigate to
the example you want to run and execute it.

```
!git clone https:// yourname:your-secret-token
@gitlab.com/nvidia/modulus/examples.git
```

In just this short amount of time, you are ready to run some of their tutorials! I have done this for several examples in Google Colab without issue. For example, if you want to navigate to their Helmholtz tutorial, then you would use the following lines of code to run the example. Again, it's only ideal for those who want to try an example or two out without much setup time, but if you want to do several projects or go more in detail, please use their recommendation installation procedure.

```
%cd examples/helmholtz/
!python helmholtz.py
```

At long last, let's talk about the super resolution example. In short, the final capability with this model is that you can take in coarse simulation results (a coarse mesh) from a simulation and then the model will 'upscale' them to that of a finer mesh result (which is presumably more accurate). The cost savings is that more coarse simulations can be run rapidly in the future and the machine learning model will be the mechanism by which a higher resolution result is generated, rather than by running very expensive and time-consuming simulations that are often prohibitively expensive in the hardware needed to conduct the simulation. In an industrial scenario, this means theoretically that cheaper simulations would be upscaled to the quality of results that ordinally would not be possible to generate on an industry deadline (not to mention you could run many of these simulations).

The architecture NVIDIA Modulus uses for this problem is shown below in Figure 31. It originates from a computer vision work [34] for an image upscaling application. The architecture has two components. The front is a series of residual blocks comprised of two standard convolutional operations. Recall that convolutional operations are things like applying filters to extract insights from the data like edges, so the front of the model is used to extract information from the input data (with residual blocks to ensure the model can be made very deep and still converge without 'vanishing gradients'). The second half of the architecture is for multiple upscaling blocks, which each have a convolutional operation, a pixel shuffle upscaling, and activation function. A pixel shuffle technique works by rearranging the elements of a tensor to increase its spatial resolution, as opposed to simply interpolating between pixels. So as far as I understand it, it is doubling that height and width of the feature maps to increase by a factor of two.

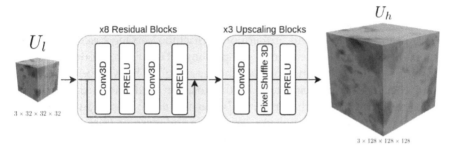

Figure 31: NVIDIA Modulus's super-resolution network, based on [34.]

What's also neat about this project is that it requires pyJHTDB; a wrapper for the Johns Hopkins University Turbulence Database Cluster library. This is a fantastic resource for accessing extremely high-quality turbulence data.

The results from this project were pretty good! If I re-visited this project and wanted to extend it, I would enjoy trying to use different cases from the JHTDB for training. Also, I would try to incorporate 0D predictions into the machine learning model to accompany the local fields predictions. For example, a skin friction coefficient value for a certain configuration. This would be very useful in industry, since while the local beautiful results are fascinating, most simulation work in industry is a means to an end for the design performance hinged on 0D scalar values (e.g. drag coefficient).

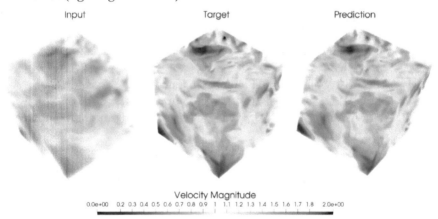

Figure 32: Velocity magnitude for a validation case using from the Modulus website for the turbulence super resolution tutorial [35].

NASA Workshop Cases

This section is one that bears several fruits for the reader; we will call out the long standing and wonderful 'turbulence modeling resource' by the NASA Langley Resource Center, and furthermore a machine learning specific resource that was the outcome of their recent symposium ("Turbulence Modeling: Roadblocks, and the Potential for Machine Learning"). The symposium was a friendly challenge for participants to try their hand at predicting a series of bench marking problems for whichever methods they wanted to employ, with certain focus in this symposium in applying data-driven and machine learning techniques. This is a great resource for us with the free datasets they provide, as well as the validation data from the turbulence modeling resource online, and the recordings from the presentations from each day in the symposium. For those trying to build up your machine learning portfolio or detect an area of research you want to pursue, these are very enabling resources!

Let's get into the different benchmarking cases from the NASA website [36]. For each case there will be one bullet point to describe the configuration and then the one below it that will state the objective for the machine learning model.

- 2D Zero Pressure Gradient Flat Plate Validation Case
  - Show (1) Cf vs. x and (2) u^+ vs. $\log(y^+)$ at x=0.97; compare with theory
- 2D Fully-Developed Channel Flow at High Reynolds Number Validation Case
  - Show u^+ vs. $\log(y^+)$ at x=500; compare with theory
- Axisymmetric Subsonic Jet
  - Show (1) u/U_{jet} vs. x/D_{jet}, (2) u/U_{jet} vs. y/D_{jet} at 5 specified stations, and (3) $u'v'/(U_{jet}^2)$ vs. y/D_{jet} at 5 specified stations; compare with experiment
- 2D NASA Wall-Mounted Hump Separated Flow Validation Case
  - Show (1) C_p vs. x/c, (2) C_f vs. x/c, (3) u/U_{inf} vs. y/c at 7 specified stations, and (4) $u'v'/(U_{inf}^2)$ vs. y/c at 7 specified stations; compare with experiment
- 2D NACA 0012 Airfoil Validation Cases (4 separate cases)
  - Angles of attack = 10, 15, 17, and 18 deg.
  - Show (1) C_L vs. alpha, (2) C_D vs. C_L, (3) C_p vs. x/c, and (4) C_f (upper surface) vs. x/c; compare with experiment (except for C_f)

Have fun!

More Datasets

This section might be a little awkward to read in a (physical) book, but every time I have a discussion with someone who wants to get into machine learning, they cherish these resources/datasets. I say 'awkward' because essentially, I hope to point you to different places on the internet for the datasets, but of course do not want to write-out the URLs for you to type by hand. Therefore, I came up with the idea to provide URLs for each resource, with a short description, as well as compile the full list onto my LinkedIn profile for your access (in the "About me" section). I think the most convenient thing for the reader is to go to my LinkedIn profile and copy/paste the URLs directly for the datasets they are interested in from their web browser. Further, if the links change over time I can update them accordingly on my LinkedIn page; so please use that as your default way for accessing them. Feel free to scan the QR code to be routed to my LinkedIn page or type the URL manually.

Justin Hodges, PhD ⊘ (He/Him)
Senior AI/ML Technical Specialist (Product Management)
Talks about #ai, #cae, #cfd, #simulation, and #machinelearning
Orlando, Florida, United States · Contact info

Siemens Digital Industries Software

Stanford Continuing Studies

URL: https://www.linkedin.com/in/justin-hodges-phd-3432a58b/

To differentiate from the above section, these are not codes or projects you can browse and experiment with, but rather datasets you can stockpile and set aside for future projects you may want to do. I went ahead and split them into two groups: datasets that are relevant to our niche world of computer aided engineering and simulation (e.g. a dataset comprised of simulation data) and more general machine learning datasets (e.g. a dataset of housing prices). In case you are new to this concept, I just want to point out that you can benefit greatly from the latter grouping for your machine learning projects in CAE applications, so do keep them in mind.

Aerospace, CAE, and Mechanical engineering specific resources:

Machine Learning for Physical Simulation Challenge [49]	
This competition aims at promoting the use of ML based surrogate models to solve physical problems, through a task addressing a CFD use case: Airfoil design	

Stanford Engineering Center for Turbulence Research [50]	
DNS statistics saved into data files in separate repositories for fully developed turbulent pipe flow, transitional flow in a pipe, and a zero-pressure-gradient flat-plate boundary layer configuration.	

MegaFlow2D: A Parametric Dataset for Machine Learning Super-resolution in Computational Fluid Dynamics Simulations [51]	
A dataset of over 2 million snapshots of parameterized 2D fluid dynamics simulations of 3000 different external flow and internal flow configurations	

Vreman Research [52]	
Databases of Direct Numerical Simulations of turbulent channel flow	

Simulation of heat transfer phenomena with supercritical CO2 [53]	
Direct numerical simulation database for supercritical dioxide.	

Johns Hopkins Turbulence Databases [54]	
This website is a portal to an Open Numerical Turbulence Laboratory that enables access to multi-Terabyte turbulence databases	

More General AI Datasets & Resources:

Kaggle [55]	
Kaggle datasets are a diverse collection of datasets made available on the Kaggle platform, encompassing a wide range of topics. Kaggle is extremely popular and hosts lots of discussion boards, coded projects, and datasets.	

UCI Machine Learning Repo [56]	
As of today, it includes 664 datasets as a service to the machine learning community. Here, you can donate and find datasets used by millions of people all around the world!	

Google Dataset Search [57]	
Dataset Search is a search engine for datasets. Using a simple keyword search, users can discover datasets hosted in thousands of repositories across the Web.	

Still lacking inspiration for which projects to pursue, or where machine learning can add value in our realm of mechanical and aerospace applications? Here is a hefty list that can help you brainstorm (and yet, there are many, many more examples).

Turbulence Modeling

- Improved turbulence modeling via ML-based turbulent diffusivity predictions [58]
- Improved turbulence modeling via ML-based turbulent Prandtl number predictions [59]
- ML-based Reynolds stressed improvements in turbulence modeling [60]
- Synthetic turbulence generator for inlet boundaries [61]
- AI-learned wall functions in turbulence modeling [62]

Turbomachinery/Aerospace

- Modeling turbulence uncertainty in compressor stall [63]
- Performance prediction and inverse design of turbocharger [64]
- Identification of losses in turbomachines [65]
- Compressor parameterized design [66]
- Cooling design correlations [67]
- Real-time deterioration of cycle components [68]
- Real-time performance modeling of aero engines [69]
- Anomaly detection of real-time operation [70]

Combustion

- Spray characterization prediction [71]
- Combustion Air/Fuel Mixing Design [72]
- Feature extraction for improved description of reacting flows
- Reducing cost of large-scale combustion modeling via dimensionality reduction
- Developing adaptive combustion closures and chemistry models [73]

9. LEARNING PATHWAY FOR BEGINNERS

Without exaggeration, I have received several hundreds of requests that go something like "what steps should I take to learn machine learning", or "how do I learn enough about AI to use it in my (CAE) day job". It's impossible to reach everybody, and part of my motivation to write this book was to scale my effort to reach more people.

This final chapter of the book seems like a great opportunity to share my baseline recommendation on what steps to take to efficiently learn machine with the end goal of being comfortable to use it in your research or industry work. The steps often start out the same for most, but then diversify from one person to the next based on the area they are interested in. But I'll touch on that soon.

Here is my high-level opinion on a practical way to learn AI/ML for a CAE engineer (since you cannot learn everything about everything). Think of it as a cone, starting from a wide base:

Step 1: Start broad. Cover a wide range of fundamental concepts in general machine learning.

Step 2: Narrow your focus to fewer topics, building up your understanding and proficiency in specific key methods/topics that really have big impact for your application (e.g. CFD, turbulence modeling, system simulation, etc.). This will involve exploring literature and different topics to identify what machine learning concepts you should be focusing on.

Step 3: Hands-on projects and experience so you can go from a general understanding of specific machine learning subjects to someone that can contribute to the field. For example, you may start out by learning that GNNs are the standard approach for working with a specific type of simulation data, but in step three you start to go from a topical understanding to getting more experience with the best practices and advanced architectures that are being used.

Step 1: Learn the basics of machine learning [~80 hours].
Pursue a broad introduction to machine learning that should help you feel more comfortable with some of the fundamental concepts. While not overly deep into any individual topic, this coverage will not shy away from basic mathematics involved and will get you started with hands-on experience on some great examples. This is not specific to mechanical and aerospace engineering applications, but rather a coverage of the basics in 'mainstream' machine learning.

I would like to highlight a few important topics that I recommend:
- Defining what machine learning is with some basic applications covered
- Overview and essential pieces in both regression and classification problems
- Supervised algorithms:
  - Linear, logistic, and polynomial regression
  - Decision trees and random forests
  - K-nearest neighbor
  - support vector machines
  - neural networks
- Dimensionality reduction techniques
  - Linear discriminant analysis
  - Principal component analysis
- Unsupervised learning algorithms
  - K-means clustering
  - Autoencoders
- Analysis tools, such as model evaluation and metrics, bias/variance, error analysis, to name a few

There is no shortage of resources available to learn AI/ML. In fact, so many that people often find it overwhelming when they first begin - "Where do I start? What's useful? What are the best courses?". While learning is a continuous process, for the basics I recommend starting with a certain course (start to finish) and then evaluate where you are. The class I recommend is taught by Andrew Ng, who has been critical in the advancement of artificial intelligence throughout his various impactful roles and affiliations (Stanford University, Google Brain, Braidu, and Landing.AI, to name a few). The course is hosted on Coursera and called "Machine Learning Specialization" [73] and will review these concepts (and more).

Once you feel that you have a grasp of the fundamentals, you can gradually shift your time towards studying CAE-specific applications of machine learning, to survey what is out there and develop a focus area that is meaningful and important to you.

Step 2: Learning the basics of machine learning applied to engineering (CAE, simulation, measurement) [1-2 months].
While there are seemingly unlimited resources which cover general AI/ML topics, there are very few which provide such an education in the CAE/simulation domain (hence in part why I felt a duty to write this book to

help change that). Of course, not all topics from mainstream AI/ML will be utilized equally in our CAE applications, so my suggestion is to narrow down your focus to the areas you need to hone your skills in.

There are some great resources out there by pioneers in the field, so let's start with calling attention to a few classes I would recommend browsing by them. I definitely don't recommend you fully consume every episode in each, but rather start identifying certain important ML topics (for your application of interest, for example turbulence modeling) and diving deeper into them. These resources cover many topics and can help you build more fundamentals due to the eloquence and wisdom the authors have.

The VKI Lecture Series – Machine Learning for Fluid Mechanics [74]. A series of lectures on YouTube that dives into reviewing the mathematical and theoretical backgrounds, with a few hands-on coding exercises, on topics ranging from data decomposition methods pioneered in fluid mechanics to somewhat modern machine learning methods. They intentionally start way-back on introductory topics but span a broad range towards more recent state of the art research methods. The path is set with the intention of covering both well-known methods that are still very useful today, as well as more recent advancements, such that students can keep up with the pace of evolution by building from existing knowledge. There is a lot of emphasis in this course on analyzing both physical and numerical data, modeling order reduction, and data-driven methods, and flow control.

Steve Brunton's YouTube channel [75]. Wildly informative and outrageously aesthetic videos by Dr. Brunton which cover a lot of topics in machine learning, data-driven methods, controls, dynamical systems, and fluid mechanics. The videos are excellent tools to use if you feel like you are struggling on a certain topic and want a deeper explanation that is not shy about showing mathematics. It is oriented for the technical student but does a fantastic job at being inclusive to those who don't have a lot of background on the subject. I frequently find myself just watching his videos for fun, but if you wanted to be efficient about consuming them then I would sort his videos by popularity and check your fundamentals on some of the most important topics to you, like his videos on neural network architectures, Principal Component Analysis, Physics Informed Machine Learning, and Machine learning for fluid mechanics (to name a few).

After you review some of these resources to refine your machine learning education to areas of importance in CAE, I recommend you start doing a good ole fashion literature review for about one month of consistent effort before going into 'step 3'.

If you have published any academic thesis before, undergraduate or graduate alike, you probably are familiar with this exercise. In short, you need to read enough literature on enough different topics so that you can identify what machine learning methods are relevant in whatever application you are currently working or interested in. Also, browsing this many papers will probably help you identify what types of machine learning projects are interesting to you in general, and which you rather not become an expert in. Even for me, as someone who has an insatiable thirst for growing myself in machine learning over the last 6 years...there are plenty of things I am not fascinated by and would rather avoid.

I would start with meta/survey academic publications on certain topics (e.g. machine learning for turbulence modeling, or PINNs for fluid mechanics, etc.). These will be comprehensive enough to review the state of the art, with some nod back to big milestones reached thanks to past seminal works. It will help expose you to a lot of areas of ongoing research, and what machine learning developments are supporting it, so you can hone your interests. Further, you will see a ton of references that you can use to start keeping a memory on what professors or research organizations are really paving the way on topics you care about. You can go to Google scholar and look them up to keep reading more about the things they've published. Further, since you cannot read every paper and should balance deeply reading some papers while only skimming others, you can deep-dive on fundamentals in a very picky way from the really famous authors that published some amazing works.

Another great (and practical) idea is to make a list of AI companies that you are aware of in your industry and look up their publications and patents (usually on their website, or at minimum back to Google scholar using their author information). At this point, I'd bet there are probably AI startups or research teams inside OEM companies in most reader's field that have made quite a buzz and possible disruption to the market. It can often be very educational to dig into reading their publications and collateral to learn what are the state-of-the-art machine learning models/approaches for the industry you are working in. There's a lot of hype out there, and this way of research can help give you a more concrete view on what's really being used in industry. Of course, not everything will be public, so you'll have to live with not getting all information about how their technologies work.

Step 3: Get hands-on experience.

As you go into step 3, you should have a reasonably good idea on what machine learning techniques you should start learning to achieve the highest fidelity capabilities in your interest area. So, in my opinion, it's time to shift into getting hands-on experience to go with the theoretical knowledge you've been gathering from reading and online resources.

Nominally, you would start take baby steps in your hands-on projects. Starting from very basic tutorials and machine learning exercises for simple to use models and datasets. Then, over time, refine your projects to more advanced exercises specific to CAE. Remember, it is much more common in AI/ML publications to publicly share the codes and datasets used, so when ready do transition to using these as cherished resources for learning.

This section is very short because much of the enablement for gaining hands-on experience was already covered in the "DATASETS & PROJECTS" chapter. So do refer back to that chapter for detailed recommendations.

A parallel step: Upskilling in Python

While the above steps assume you begin your machine learning self-education journey with some basic Python knowledge, we should certainly have more discussion on this. It is interesting because you certainly do not need to be an expert in Python to begin working on machine learning projects, but you definitely have to start your journey with some base knowledge, and you should prioritize ramping-up your proficiency in Python along the way. I am a fan of learning more-and-more Python as you go through machine learning curriculums, rather than frontloading a long period of time focused purely on Python education (you'll spend a lot of time learning things that aren't essential to your machine learning project work that way).

After some personal communication with Maksym Kalaidov [76], who has great content on learning python as a CAE engineer interested in machine learning, I am delighted to have received his permission to include his figure here, as well as some comments from our personal communication. It's easy to just say 'go learn Python', but in current times that so overwhelming with the vast number of resources available and considering that not all is relevant to our ambition of AI in CAE.

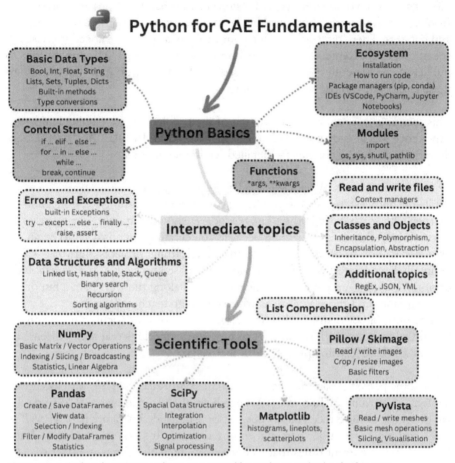

Figure 33: A Python learning roadmap suggested by Maksym Kalaidov [76] in CAE applications!

In my view, the blue boxes are things you must understand before trying out any machine learning project or code. If you cannot setup your space to code your project (even a simple one), with the necessary libraries/versions/ecosystem, then it simply won't run.

While the yellow boxes are things you can be mediocre in to start, I think you can still move forward to begin your machine learning projects anyway and learn as you go. Even with only a mediocre knowledge, you'll probably have a reasonable idea on what you should be doing with your project/code, but you just may need to search a few specific things on the internet here-and-there to correct your code. For example, you may know how to read and write data when working with some formats (e.g. csv) but you might have to search for the appropriate syntax when using others (e.g. cgns).

Similarly, I think some mediocre knowledge for orange boxes is all you need before you can be ready to begin your machine learning projects. Some libraries, like NumPy and Pandas, you will use a lot in your projects for very basic and core operations which you cannot avoid (managing your data). So, a basic idea on how to use them at the start is sufficient, but undoubtedly you will continue to grow more familiar with different aspects of each and need to learn on the fly. For example, you will learn more ways to transform your data with Pandas commands, as needed in your different projects. This shouldn't stop you from getting started though, as even experts are still learning when projects pitch some unique challenge you'll need to tackle via these libraries.

So now that you have some list of key terms to touch on, where should you turn to gather that education? Well, there are so many resources available, it is hard to narrow it down. It is okay to use one that you like and works well for you, even if it's not popular. But my baseline suggestion is Harvard's CS50's Introduction to Programming with Python. It's free and exceptional!

Lastly, a few tips: pick an editor that is easy to setup and stick with it over time, don't copy and paste code you don't understand, and try to stay working in standard libraries as much as you can. And finally, especially knowing how engineers are; don't overthink how to get started. Don't delay getting started by weeks from worrying too much about which course is best to learn from, or which version of some library is ideal to start with, just don't wait another day to get started and remain consistent – you will be better at writing code by the end of it.

REFERENCES

[1] Hodges, Justin, et al. "Application of Machine Learning and CFD to Model the Flow in an Internal Combustion Engine". International CAE Conference. 2021.

[2] Franklin, Gene F., et al. Feedback control of dynamic systems. Vol. 4. Upper Saddle River: Prentice hall, 2002.

[3] Julia Elliott, Paul Mooney. (2021). 2021 Kaggle Machine Learning & Data Science Survey. Kaggle. https://kaggle.com/competitions/kaggle-survey-20

[4] Gross, Jurgen, and Jürgen Groß. *Linear regression*. Vol. 175. Springer Science & Business Media, 2003.

[5] Deng, L. (2012). The mnist database of handwritten digit images for machine learning research. IEEE Signal Processing Magazine, 29(6), 141–142.

[6] Wach, Noah L., et al. "Data re-uploading with a single qudit." *arXiv preprint arXiv:2302.13932* (2023).

[7] Hodges, Justin et al. "Artificial Intelligence and Machine Learning for Ship Design". Proceedings ICASS, September 2022.

[8] Bergman, Theodore L., et al. *Introduction to heat transfer*. John Wiley & Sons, 2011.

[9] Pedregosa, Fabian, et al. "Scikit-learn: Machine learning in Python." *the Journal of machine Learning research* 12 (2011): 2825-2830.

[10] Monolith. "AI Software: Engineering Product Development." Monolith, 2022, https://www.monolithai.com/.

[11] Kokhlikyan, Narine, et al. "Captum: A unified and generic model interpretability library for PyTorch." arXiv preprint arXiv:2009.07896 (2020).

[12] Pedregosa, Fabian, et al. "Scikit-learn: Machine learning in Python." the Journal of machine Learning research 12 (2011): 2825-2830.

[13] Lecun Y, Bottou L, Bengio Y, Haffner P (1998Nov) Gradient-based learning applied to documentrecognition. Proc IEEE 86(11):2278–2324

[14] Open Source Imaging Consortium (OSIC). (2020). OSIC Pulmonary Fibrosis Progression. Kaggle. https://www.kaggle.com/competitions/osic-

pulmonary-fibrosis-progression/code

[15] TensorFlow. "EfficientNetB5." TensorFlow, Year of access, https://www.tensorflow.org/api_docs/python/tf/keras/applications/efficientn et/EfficientNetB5.

[16] Brownlee, J. (2023). How to Create Your First Neural Network in Python with Keras. Machine Learning Mastery. https://machinelearningmastery.com/tutorial-first-neural-network-python-keras/

[17] Akkari, Nissrine, et al. "A bayesian nonlinear reduced order modeling using variational autoencoders." *Fluids* 7.10 (2022): 334.

[18] Solera-Rico, Alberto, et al. "β-Variational autoencoders and transformers for reduced-order modelling of fluid flows." *arXiv preprint arXiv:2304.03571* (2023).

[19] Wu, Rundi, Chang Xiao, and Changxi Zheng. "Deepcad: A deep generative network for computer-aided design models." Proceedings of the IEEE/CVF International Conference on Computer Vision. 2021.

[20] Chung, Wai Tong, et al. "BLASTNet: A call for community-involved big data in combustion machine learning." *Applications in Energy and Combustion Science* 12 (2022): 100087.

[21] Sanchez-Lengeling, Benjamin, et al. "A Gentle Introduction to Graph Neural Networks." *Distill*, vol. 6, no. 8, 17 Aug. 2021, https://doi.org/10.23915/distill.00033.

[22] Portal-Porras, Koldo, et al. "Hybrid LSTM+ CNN architecture for unsteady flow prediction." *Materials Today Communications* 35 (2023): 106281.

[23] Hou, Yuqing, et al. "A novel deep U-Net-LSTM framework for time-sequenced hydrodynamics prediction of the SUBOFF AFF-8." *Engineering Applications of Computational Fluid Mechanics* 16.1 (2022): 630-645.

[24] Zahn, Rebecca, et al. "Application of a long short-term memory neural network for modeling transonic buffet aerodynamics." *Aerospace Science and Technology* 113 (2021): 106652.

[25] Hendrickx, Kilian, et al. "A general anomaly detection framework for fleet-based condition monitoring of machines." *Mechanical Systems and Signal Processing* 139 (2020): 106585.

[26] ElSaid, AbdElRahman, et al. "Using LSTM recurrent neural networks to predict excess vibration events in aircraft engines." *2016 IEEE 12th International Conference on e-Science (e-Science).* IEEE, 2016.

[27] Sierra-Garcia, J. Enrique, and Matilde Santos. "Deep learning and fuzzy logic to implement a hybrid wind turbine pitch control." *Neural Computing and Applications* (2021): 1-15.

[28] Grinsztajn, Léo, Edouard Oyallon, and Gaël Varoquaux. "Why do tree-based models still outperform deep learning on typical tabular data?." *Advances in Neural Information Processing Systems* 35 (2022): 507-520.

[29] Mathan. Profile on Kaggle. Kaggle, https://www.kaggle.com/mathan. Accessed 21 December 2024.

[30] Ling, J., and Templeton, J., 2015. "Evaluation of Machine Learning Algorithms for Prediction of Regions of High Reynolds Averaged Navier Stokes Uncertainty". Physics of Fluids, 27.

[31] Gorman, R. P., and Sejnowski, T. J. (1988). "Analysis of Hidden Units in a Layered Network Trained to Classify Sonar Targets" in Neural Networks, Vol. 1, pp. 75-89.

[32] ProsperityAI. (n.d.). Ensemble-learning: ensemble_methods.ipynb. GitHub. Retrieved November 1, 2023 from https://github.com/prosperityai/ensemble-learning/blob/master/ensemble_methods.ipynb

[33] Géron, Aurélien. *Hands-on machine learning with Scikit-Learn, Keras, and TensorFlow.* " O'Reilly Media, Inc.", 2022.

[34] Ledig, Christian, et al. "Photo-Realistic single image super-resolution using a generative adversarial network" Conference on Computer Vision and Pattern Recognition, 2017

[35] NVIDIA Modulus Sym User Guide: Turbulence Super Resolution." NVIDIA, 1 Dec. 2023, https://docs.nvidia.com/deeplearning/modulus/modulus-sym/user_guide/intermediate/turbulence_super_resolution.html.

[36] "Workshop on Turbulence Modeling for Propulsion and Power Systems." NASA Langley Research Center, 2022, https://turbmodels.larc.nasa.gov/turb-prs2022.html. Accessed 15 Dec. 2023.

[37] "BlastNet." BlastNet, https://blastnet.github.io/index.html. Accessed 15 Nov. 2023.

[38] Chung, Waitong. "Smallhit-TFMultigpu." Kaggle, https://www.kaggle.com/code/waitongchung/smallhit-tfmultigpu. Accessed 15 Nov. 2023.

[39] cfl-minds. "gnn_laminar_flow." *GitHub*, github.com/cfl-minds/gnn_laminar_flow/tree/main. Accessed October 2023.

[40] Chen, Junfeng, Elie Hachem, and Jonathan Viquerat. "Graph neural networks for laminar flow prediction around random two-dimensional shapes." Physics of Fluids 33.12 (2021).

[41] cfl-minds. "cnn_drag_prediction." *GitHub*, github.com/cfl-minds/cnn_drag_prediction. Accessed October 2023.

[42] Moseley, Ben. "So What Is a Physics-Informed Neural Network?" Ben Moseley's Blog, https://benmoseley.blog/my-research/so-what-is-a-physics-informed-neural-network/. Accessed Jan. 2024.

[43] Hodges, Justin. "Investigating Film Cooling Flows with Advanced Turbulence Modeling, Machine Learning, and Experimental Methods." (2020).

[44] Ames, F. E., L. A. Dvorak, and M. J. Morrow. "Turbulent augmentation of internal convection over pins in staggered-pin fin arrays." J. Turbomach. 127.1 (2005): 183-190.

[45] Wang, Jian-Xun, et al. "A comprehensive physics-informed machine learning framework for predictive turbulence modeling." arXiv preprint arXiv:1701.07102 (2017).

[46] Ling, Julia, and Jeremy Templeton. "Evaluation of machine learning algorithms for prediction of regions of high Reynolds averaged Navier Stokes uncertainty." Physics of Fluids 27.8 (2015).

[47] Milani, Pedro M., Julia Ling, and John K. Eaton. "Physical interpretation of machine learning models applied to film cooling flows." Journal of Turbomachinery 141.1 (2019): 011004.

[48] Wang, Jian-Xun, Jin-Long Wu, and Heng Xiao. "Physics-informed machine learning approach for reconstructing Reynolds stress modeling discrepancies based on DNS data." Physical Review Fluids 2.3 (2017): 034603.

[49] "ML for Physical Simulation Challenge." IRT SystemX, accessed 15 February 2024, https://ml-for-physical-simulation-challenge.irt-systemx.fr/.

[50] "Research Data." Center for Turbulence Research, Stanford University, accessed 15 February 2024, https://ctr.stanford.edu/about-center-turbulence-research/research-data.

[51] Xu, Wenzhuo, Noelia Grande Gutierrez, and Christopher McComb. "MegaFlow2D: A parametric dataset for machine learning super-resolution in computational fluid dynamics simulations." Proceedings of Cyber-Physical Systems and Internet of Things Week 2023. 2023. 100-104.

[52] Vreman, Bert. "Channel Flow." Vreman Research, accessed 15 February 2024, http://www.vremanresearch.nl/channel.html.

[53] "Institute of Combustion and Power Plant Technology: Direct Numerical Simulations (DNS) of sCO2 Turbulence." University of Stuttgart, accessed 15 February 2024, https://www.ike.uni-stuttgart.de/en/research/sco2/dns/.

[54] Kanov, Kalin, et al. "The Johns Hopkins turbulence databases: an open simulation laboratory for turbulence research." Computing in Science & Engineering 17.5 (2015): 10-17.

[55] "Datasets." Kaggle, accessed 15 February 2024, https://www.kaggle.com/datasets.

[56] Asuncion, Arthur, and David Newman. "UCI machine learning repository." (2007).

[57] "Google Dataset Search." Google, accessed 8 January 2024, https://datasetsearch.research.google.com/.

[58] Ellis, Christopher D., Hao Xia, and Gary J. Page. "LES informed data-driven modelling of a spatially varying turbulent diffusivity coefficient in film cooling flows." Turbo Expo: Power for Land, Sea, and Air. Vol. 84171. American Society of Mechanical Engineers, 2020.

[59] Costa, Fabíola Paula, et al. "Evaluation of a Machine Learning Turbulence Model in a Square Transverse Jet in Crossflow." Turbo Expo: Power for Land, Sea, and Air. Vol. 84171. American Society of Mechanical Engineers, 2020.

[60] Akolekar, Harshal D., et al. "Integration of machine learning and computational fluid dynamics to develop turbulence models for improved low-pressure turbine wake mixing prediction." Journal of Turbomachinery 143.12

(2021): 121001.

[61] Corsini, Alessandro, et al. "Machine learnt synthetic turbulence for LES inflow conditions." Turbo Expo: Power for Land, Sea, and Air. Vol. 84058. American Society of Mechanical Engineers, 2020.

[62] Tieghi, Lorenzo, et al. "A Machine-Learnt Wall Function for Rotating Diffusers." Journal of Turbomachinery 143.8 (2021): 081012.

[63] He, Xiao, Fanzhou Zhao, and Mehdi Vahdati. "Uncertainty quantification of Spalart–Allmaras turbulence model coefficients for compressor stall." Journal of Turbomachinery 143.8 (2021): 081007.

[64] Thatte, Azam, et al. "An Artificial Intelligence Based Method for Performance Prediction and Inverse Design of Hydraulic Turbochargers." Turbo Expo: Power for Land, Sea, and Air. Vol. 84195. American Society of Mechanical Engineers, 2020.

[65] Angelini, Gino, et al. "Identification of losses in turbomachinery with machine learning." Turbo Expo: Power for Land, Sea, and Air. Vol. 84058. American Society of Mechanical Engineers, 2020.

[66] Goldenberg, Vlad, et al. "A Numerical Approach to Centrifugal Compressor Stage Flow Path Design Synthesis and Optimization." Turbo Expo: Power for Land, Sea, and Air. Vol. 84089. American Society of Mechanical Engineers, 2020.

[67] Yang, Li, Qi Wang, and Yu Rao. "Modeling Superposition of Flat Plate Film Cooling Under Complicated Conditions Using Recurrent Neural Networks." Turbo Expo: Power for Land, Sea, and Air. Vol. 84171. American Society of Mechanical Engineers, 2020.

[68] Yang, Shu-Bo, et al. "A Self-Tuning Model Framework Using K-Nearest Neighbors Algorithm." Turbo Expo: Power for Land, Sea, and Air. Vol. 84140. American Society of Mechanical Engineers, 2020.

[69] Ibrahem, Ibrahem MA, et al. "An ensemble of recurrent neural networks for real time performance modeling of three-spool aero-derivative gas turbine engine." Journal of Engineering for Gas Turbines and Power 143.10 (2021): 101004.

[70] Goyal, Vipul, et al. "Prediction of gas turbine performance using machine learning methods." Turbo Expo: Power for Land, Sea, and Air. Vol. 84157. American Society of Mechanical Engineers, 2020.

[71] Chaussonnet, Geoffroy, et al. "Best architecture of an artificial neural network to model prefilming airblast atomization: Not so deep learning." Journal of Engineering for Gas Turbines and Power 143.7 (2021): 071006.

[72] Li, Suhui, et al. "Prediction of the Autoignition of a Fuel Jet in a Confined Turbulent Hot Coflow Using Machine Learning Methods." Turbo Expo: Power for Land, Sea, and Air. Vol. 84126. American Society of Mechanical Engineers, 2020.

[73] "Machine Learning Introduction." Coursera, accessed 15 February 2023, https://www.coursera.org/specializations/machine-learning-introduction#courses.

[74] Von Karman Institute for Fluid Dynamics. "The VKI Lecture Series – Machine Learning for Fluid Mechanics." YouTube, uploaded by VKI, 3 March 2023, https://www.youtube.com/watch?v=Xq7ROz97lbk&list=PLrNYZCDknzQcg_UiXoMVPm0M-Kh7ZTpQ_.

[75] Brunton, Steve. "Eigensteve." YouTube, https://www.youtube.com/@Eigensteve. Accessed 3 March 2023.

[76] Kalaidov, Maksym. LinkedIn Profile. LinkedIn, https://www.linkedin.com/in/maksym-kalaidov/. Accessed 12 March 2023.

Made in United States
Troutdale, OR
12/10/2024

26143826R00096